Local and Urban Governance

Series Editor
Carlos Nunes Silva, Institute of Geography and Spatial Planning, University of Lisbon, Lisbon, Portugal

This series contains research studies with policy relevance in the field of sub-national territorial governance, at the micro, local and regional levels, as well as on its connections with national and supranational tiers. The series is multidisciplinary and brings together innovative research from different areas within the Social Sciences and Humanities. The series is open for theoretical, methodological and empirical ground breaking contributions. Books included in this series explore the new modes of territorial governance, new perspectives and new research methodologies. The aim is to present advances in Governance Studies to scholars and researchers in universities and research organizations, and to policy makers worldwide. The series includes monographs, edited volumes and textbooks. Book proposals and final manuscripts are peer-reviewed.

The areas covered in the series include but are not limited to the following subjects:

- Local and regional government
- Urban and metropolitan governance
- Multi-level territorial governance
- Post-colonial local governance
- Municipal merger reforms
- Inter-municipal cooperation
- Decentralized cooperation
- Governance of spatial planning
- Strategic spatial planning
- Citizen participation in local policies
- Local governance, spatial justice and the right to the city
- Local public services
- Local economic development policies
- Entrepreneurialism and municipal public enterprises
- Local government finance
- Local government and sustainable development
- Anthropocene and green local governance
- Climate change and local governance
- Smart local governance

The series is intended for geographers, planners, political scientists, sociologists, lawyers, historians, urban anthropologists and economists. **Local and Urban Governance—now indexed in Scopus**

Łukasz Damurski

The Foundations of Multi-channel Neighborhood Governance

Springer

Łukasz Damurski
Faculty of Architecture
Wrocław University of Science and Technology
Wrocław, Poland

ISSN 2524-5449 ISSN 2524-5457 (electronic)
Local and Urban Governance
ISBN 978-3-031-75126-4 ISBN 978-3-031-75124-0 (eBook)
https://doi.org/10.1007/978-3-031-75124-0

This Springer imprint is published by the registered company Springer Nature Switzerland AG
The registered company address is: Gewerbestrasse 11, 6330 Cham, Switzerland

The smart city is not just alive, but live. On the surface, its urbanism may appear alike to any other in the present (…). What remains largely unseen is the very promise of ubiquitous computing to create a networked layer that mediates these interactions over the course of seconds and seasons. If designers are to gain meaningful agency within this framework, we need to meet it on its own terms and begin to adapt our practices to become similarly continuous and contingent.

Belesky, P., 2020. A perennial practice. Designing between urban landscape and urban network. "Urban futures". Special issue: Designing the Digitalised City, pp. 100–107.

The metaverse is one of the technologies of the future that is generating potential exodus of people from real places to digital ones.

Oleksy, T., Wnuk, A. i Piskorska, M., 2023. Migration to the metaverse and its predictors: Attachment to virtual places and metaverse-related threat. Computers in Human Behavior, Volume 141(2023, 107642), pp. 1–10.

The inevitable interactions among technology development, social trends, policy responses, and public management adaptations will continue to provide a dynamic field for learning and action. Many different forms of ICT-enabled governance are possible (…) for weaving salient concepts and tools more usefully into the broad purposes, functions, and performance of government.

Dawes, S. S., 2008. The evolution and continuing challenges of e-governance. Public administration review, 68(2008), pp. 86–S102.

Preface

There are many reasons to study neighborhoods. As an inherent part of human settlement strategies, they have evolved throughout history, following the ever-changing patterns of civilization, and reflecting the values and priorities of individuals, communities, and societies in successive epochs. Thus neighborhoods offer a reliable insight into social relations, land-use changes, and economic development in any habitable place of the world.

But neighborhood development is not only a rewarding object of research, with high cognitive values of its own. It is also subject to management and governance, requiring appropriate policy approach, answering its current challenges and supporting its long-lasting endurance. Today (in 2020s) it seems that neighborhoods find themselves in a turning point of their history. This is not to say that their very existence is endangered with some unexpected turmoil, but rather to identify some of the risks they face and the contingency plans they can undertake. The continuous development of ICT, the expansion of online markets, changes in customer behavior, the digitalization of social interactions, democratization (to name only the most obvious ones) have a significant impact on the spatial and functional neighborhood structures. Finding appropriate, relevant, and realistic responses to those phenomena in neighborhood planning is the core goal of this book.

When writing the first version of the manuscript, I initially thought of online services as the main threat that neighborhoods will have to tackle in the near future in order to survive. More and more traditional urban functions (such as shops, banks, restaurants, administration, or even groceries) are moving to the Internet, weakening the territorial basis of neighborhood livability. What about the local quality of life?—I asked myself. What about public spaces and the many roles they play in community building? How can we, as planners, respond to the online-offline competition?

Then, after several years of conducting various research projects I realized that services sector is not divided into "bricks-and-mortar" and "virtual"; that there are different things that customers do online and different things they do offline. Therefore, neighborhood governance cannot only focus on strengthening the on-site services sector but must also take action also in the online channel.

Another lesson was learned when the 2020/2021 lockdowns were introduced around the world and all the administrative processes had to be carried out remotely. A neighborhood planner in Sommerville, MA, told me how successful they had been in conducting online consultations with their communities. This made me think that it is not just the content of planning that needs to be considered offline and online: we also need to rethink the form of planning procedures. And of course, e-participation has been developing for two decades now, so we already have the necessary administrative regulations regarding online citizen involvement and a lot of good practice. What is missing, however, is a comprehensive framework that allows a hybrid approach to both the object of planning and the means of planning.

This is how the concept of multi-channel neighborhood governance was born.

Wrocław, Poland Łukasz Damurski

Acknowledgments

This book is a result of my research conducted within several research projects. Its origins stem back to 2012–2015 when I realized a grant from the National Science Centre Poland titled "Public communication in spatial planning. Redefining the role of the planner in spatial decision making process" at the Faculty of Architecture, Wroclaw University of Technology (grant number 2011/03/D/HS4/00806, further referred as NCN research project 2012–2015). It gave me the wide perspective on citizen participation theory and practice, including online tools for involving citizens.

The second research project was also funded by the National Science Centre Poland and was titled "Model of the local service centre as a tool for enhancing territorial cohesion of urban areas." It was realized at the Faculty of Architecture, Wrocław University of Science and Technology in the years 2016–2020 (grant number 2015/19/B/HS4/01301, further referred as NCN research project 2016–2020) and resulted in large amount of comparative data from several case studies across Europe. It gave me an opportunity to widen my knowledge on local (neighborhood) service centers, including the recent development of online services.

A real milestone on this research path was the Senior Award scholarship titled "Online conversion of services and its impact on neighborhood planning," obtained from the Polish-US Fulbright Commission and carried out at the Harvard University Graduate School of Design (USA) in 2021 (grant number PL/2020/38/SR, further referred to as Fulbright Senior Award scholarship 2021). It gave me a unique opportunity to test my knowledge in the American context by conducting both theoretical and empirical research in selected neighborhoods in the Boston metropolitan area. Many of the examples presented in this book are the result of this research project.

The current research grant realized within the H2020 program, called "DEMOTEC. Democratising Territorial Cohesion: Experimenting with deliberative citizen engagement and participatory budgeting in European regional and urban policies" (grant number 962553, further referred as H2020 DEMOTEC research project 2022), plays a complementary role in my research agenda. It has spatial and social references and it involves both online and offline tools for urban decision making.

But most of all this publication would not be possible without my direct intellectual interaction and cooperation with particular people. I would like to express my gratitude, especially to the following persons who supported me in writing this book:

- Katarzyna Damurska, my wife, who encouraged and motivated me with love and patience (and took care of all the rest of our life to make this project possible)
- Rick Peiser, professor at the Harvard University Graduate School of Design, Cambridge, MA, who supported me in conducting my Fulbright scholarship and proposed very relevant case study areas
- Hans Thor Andersen, research director at the Danish Building Research Institute, Aalborg University, Copenhagen, Denmark, who has been always very supportive in all my research activities
- Calos Nunes Silva, auxiliary professor at the University of Lisbon Institute of Geography and Spatial Planning, Portugal, who accompanied me and invited me to various organizational activities from the very beginning of my research career, in particular regarding my interest in e-planning and online citizen participation
- Mary Rauktis, associate professor at the University of Pittsburgh School of Social Work, Pittsburgh, PA, and her colleagues, who inspired me to continue my research on neighborhoods taking an international perspective
- Carlos Mendez, principal research fellow at the European Policies Research Centre, University of Strathclyde, Glasgow, UK, who trusted me although he had never met me before, and involved me in the H2020 DEMOTEC research project
- Eugeniusz Bagiński (1930–2020), professor emeritus at the Faculty of Architecture, Wrocław University of Science and Technology, Poland, my first mentor, tutor, and supervisor, who gave me a unique opportunity to become a researcher in urban planning and governance
- Anna Bazan-Krzywoszańska for her cordial and instant support in my application for the Excellent Science grant from the Polish Ministry of Science and Higher Education.

This project has been co-funded from the Polish state budget by the Republic of Poland Minister of Science within the Excellent Science II program (grant number MONOG/SP/0031/2024/02).

Minister of Science and Higher Education
Republic of Poland

Contents

About the Author

Łukasz Damurski is an urban planner, researcher, and academic teacher at the Faculty of Architecture, Wrocław University of Science and Technology, Poland (since 2006). In 2022 he was appointed the head of the Chair of Urban Planning and Spatial Management at FA WUST. His research interests focus on local (neighborhood) service centers, the relationship between online and offline services as well as public communication, and citizen participation in urban planning. He has played the role of the principal investigator in several international research projects funded by the European Commission, the National Science Centre Poland, and the National Centre for Research and Development. In 2021 he was awarded the Polish-US Fulbright Commission Senior Award and conducted a research project on online conversion of urban services at the Harvard University Graduate School of Design (Cambridge, MA).

Email: lukasz.damurski@pwr.edu.pl

Chapter 1
Introduction

Abstract Linking various phenomena observable in the current urban development enables drawing a general overview of the condition of contemporary neighborhoods. The immersive presence of communication technologies in urban life, virtualization of services sector, and popularization of a participatory approach in decision-making are the main reasons for conducting research on multichannel neighborhood governance. This chapter points to the gaps in the state-of-the-art, as well as arising questions connected to the insufficient appreciation of the online channel in urban governance. To ensure academic integrity and provide reliable results, every research needs to be properly structured and embedded in the background knowledge. Therefore also the materials and methods used in this book are outlined below.

1.1 Reasons for Studying Neighborhoods in a Multichannel Perspective

In recent years, there has been a growing interest in the neighborhood as a site or space for urban and social activity and in particular for governance practice. Urban policymakers around the world widely recognize the neighborhood as a fundamental, essential component of urban structure. If asked to identify its core characteristics, they would probably list functional self-containment, dominating residential land use accompanied by some everyday services (local shops, schools, health centers, sports facilities, etc.); they would point to its relatively small scale and high accessibility for pedestrians; they might also mention the values of the local community such as enjoyable, safe social life and active citizen participation in public decision-making.

However, only some of them (supposedly the younger ones) would think of the virtual representation of neighborhood, its digital presence, and online activity of its residents. This growing digital component of the neighborhood does not neglect all

Ł. Damurski, *The Foundations of Multi-channel Neighborhood Governance*, Local and Urban Governance, https://doi.org/10.1007/978-3-031-75124-0_1

its physical and social values—conversely, it rather strengthens its potential to build a comfortable living environment.

The only problem is that online neighborhood components have not been fairly conceptualized and acknowledged by the urban policymakers yet. Even though policy development has moved forward on several fronts, new phenomena continually add to an increasingly complex set of concerns which does not facilitate building holistic policy programs. The growing complexity of urban reality (cf. Paddison 2001; Wallin 2013) is fostered by various factors, and one of them is undoubtedly the dynamic development in the ICT sector (information and communications technologies) or more broadly telematics (telecommunication, transportation, informatics, and electronics) (Sassen 2010; Dawes 2008). The resulting challenges emerging in neighborhoods and other functional areas require an appropriate and effective response from the urban governance.

And of course, some aspects of virtualization have already been thoroughly studied and incorporated into the urban discourse. Services sector seems to be a clear example: the share of online shopping is constantly growing, reducing the need for in-person interaction and thus affecting the composition of retail centers; banking and public administration are also converting online very fast, and many of their brick-and-mortar premises are being locked down.

Despite the growing body of knowledge on the online conversion of various urban functions, it still has not been addressed with appropriate conceptual framework at the neighborhood level. Moreover, local urban governance seems to be reluctant in taking action in the online channel; probably due to the lack of relevant and comprehensive intellectual background.

Practitioner's Perspective
The planner's professional habits

The way that you're supposed to work is when something happens, you study it, and then you react to change your approach. There has not been enough study of this phenomenon [online conversion of services] (...). Right now, we're just doing what we always did because of inertia, based on our current understanding of the situation.

Jason King, AICP, Principal, Dover, Kohl & Partners Town Planning [interview conducted via MS Teams 11/22/2021; reviewed and authorized by the interviewee on 11/19/2021; origin: Fulbright Senior Award scholarship 2021]

Multichannel approach emerges to be an obvious follow-up of the current developments in human civilization. The urban "space of flows" (Castells 1998) embraces all dimensions of our everyday lives, starting from communication with our families, through our work to shopping and services fulfilment: all of those activities are performed both online and offline. Hence, there is an urgent need to provide some kind of holistic approach, overcoming the narrow perspective of urban governance and involving all the aspects of contemporary urban performance in decision-making processes. This is the main reason for undertaking the challenge of multi-channel governance in this book.

But one might ask: do we really need to revise the governance paradigms? After all, online services in their very essence work exactly like the traditional ones do: the providers aim to maximize their revenues while the customers want to satisfy their needs. And participatory planning has been using the online channel in its practices since the early 2000s becoming an integral part of urban governance. Yes, that is all true. However, the form of services provision has recently undergone a rapid change which has a profound impact on customer behavior and affects the spatial and functional patterns of urban areas. Online citizen participation has also developed highly advanced, sophisticated ICT tools which challenge the to-date patterns and frameworks. If policymakers, supported by researchers, are to effectively manage contemporary urban structures, they must recognize the growing role of online channel. This is how Belesky (2020) puts it: "On the surface, its urbanism may appear alike to any other in the present (…). What remains largely unseen is the very promise of ubiquitous computing to create a networked layer that mediates these interactions over the course of seconds and seasons. If designers are to gain meaningful agency within this framework, we need to meet it on its own terms and begin to adapt our practices to become similarly continuous and contingent" (Belesky 2020, p. 102).

This is an exciting time for scholars as they can guide governance practitioners and transform decision-making practices toward the forthcoming digital age (cf. Yoo et al. 2023). Urban governance (and neighborhood planning), involving its best anticipatory powers and forecasting methods, has to respond immediately to those challenges in advance with appropriate paradigms.

But how to follow the rapid changes in economy and society? How to catch up with the revolutionary online channels? Well, the first answer would be to conduct in-depth research, collect various data, construct a diagnosis of the contemporary urban areas, and provide well-informed recommendations for urban development. Paradoxically, as a researcher, I am skeptical about the results of such studies as a source of up-to-date knowledge. A typical high-quality research process takes usually several years, and its dissemination among practitioners may take another decade. This is not an effective way of providing current data and providing relevant solutions to them. Alternatively, we can use a simple but well-established method of critical, attentive (and sometimes also participatory) observation which offers an updated information about the state of urban environment.[1] This method can be complemented by information from other sources, juxtaposed and linked to draw indicative conclusions for policy practice. This approach—despite its superficial and selective character—proves to be relatively efficient in defining the current state of urban areas and anticipating the near-future trends.

[1] One of my supervisors—professor Eugeniusz Bagiński (1930–2020)—used to repeat that that observation is the elementary research method.

1.2 Research Questions and Objectives of the Book

The main ambition of this book is to contribute to urban governance paradigms by reviewing their service-related components and participatory principles in the light of the latest trends in ICT development, particularly the emerging functionalities of online channels. The main research questions are as follows:

- What are the main challenges posed by the growing presence of ICT in the urban environment? How is virtual reality changing the spatial, social, and economic structures of urban areas?
- How is urban governance responding to the challenges of virtualization? Are the existing governance paradigms relevant to the current situation at the neighborhood level?
- Should online and offline channels be included in neighborhood governance? If so, what should such multichannel governance include?

Some of those questions have already been answered by urbanism studies at the citywide scale. In particular, it has been observed that despite the online conversion effects on lifestyle, work, and services provision (even in the post-COVID-19 pandemic epoch), urban cores remain remarkably vital, which is mainly due to the advantages of proximity: the growing usage of the ICT paradoxically leads to an increased need for facilitated face-to-face contact which is necessary for information exchange, learning, and creativity. Taking the urban agglomeration perspective, it is argued that urban areas still provide unbeatable localization values for innovation and growth (Shearmur 2010).

The central thesis of this book is that those observations do not apply to the local scale and that if neighborhood governance is to fulfil the vision of sustainable living environment for contemporary urban communities, it needs to get actively involved in shaping the local services market using both online and offline channels. I will try to prove this assertion by building a particular logic of argumentation, starting with revising the urban governance paradigms, followed by characteristics of online and offline services channels, searching for urban policy answers on the current developments, and eventually by concluding with a multichannel neighborhood governance concept.

1.3 The Cognitive Sequence of the Book

The capacity to sequence information is central to any research performance. Cognitive sequencing forms the foundation stone for higher order cognition necessary for building new knowledge. It can be viewed as the ability to execute a set of actions that follow a particular order. Consistent sequential ordering (following adjacent steps or study tasks) proves to be highly effective in goal-directed activities (Savalia et al. 2016).

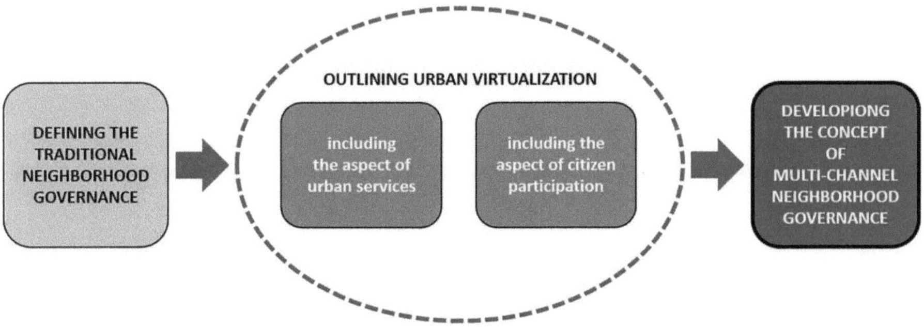

Fig. 1.1 Cognitive sequence of the book. (Source: Author's own research)

This approach has been adopted in this book for structuring the process of collecting information, synthetizing the current knowledge from various disciplines, and eventually building a new conceptual framework for neighborhood governance (Fig. 1.1). The scheme shows how starting from the description of the state-of-the-art and studying selected novel phenomena I will proceed with the novel concept of multichannel neighborhood development.

The proposed cognitive sequence has particular limitations. Even if it looks to be universal and holistic, it is in fact narrowed to particular local urban contexts and includes only two aspects of the virtualization. However, the limited range may be also its asset: the sequence is manageable and provides particular knowledge in clearly defined cognitive frames. In the following subsection, a detailed scope of the book is outlined.

1.4 The Scope of the Book

The innovations brought about by globalization encompass the entire planet, but their effects are most visible in local communities (cf. Jałowiecki and Szczepański 2002). Following this observation, the presented research—both in its theoretical and empirical components—will focus on the neighborhood scale, where the so-called glocalization pressures, reflecting the tension between traditional values and postmodern trends (Walton 2000), are clearly visible, generating a growing demand for policy action. The current dynamics of the "real" and "virtual" environments are reshaping the social and spatial behavior of local communities (Smagacz-Poziemska 2015) and posing new challenges for urban governance.

Saying this, I need to note that local communities are the first and the main subject of all public policies. A shift in urban planning observed in Europe since the 1990s proves that after decades of policies focused on structural challenges and large-scale projects, local authorities pay more and more attention to neighborhoods. This is where urban governance starts, and this is where new, dedicated, and customized approaches should be adopted.

The proposed synthesis and revision of the state-of-the-art in urban development and neighborhood governance will follow two main axes:

1. Virtualization of services sector, its impact on urban communities, and land-use structures
2. Changing decision-making patterns in local public policy as a response to development of online communication channels

The time span of the book includes particularly the last two decades (since the early 2000s through 2020s). This is the period when the Internet has expanded, changing all the levels of human everyday life and visibly affecting the performance of urban areas. In some cases, earlier phenomena are referred, forming a context for the current developments.

The geographical range of the presented research encompasses various studies as well as examples of policy practice from Europe and the USA. Such a wide scope of the research enables juxtaposing and comparing different contexts and as a result drawing a comprehensive, universal picture of the current urban development across the Western world.

1.5 Intended Audience

This book is addressed to both theoreticians and practitioners in urban governance. On one hand, it is expected to collect and synthetize the core knowledge from various disciplines related to the services sector and to local participatory decision-making. On the other hand, it provides particular solutions and inspiration for decision-making processes. As a result, it contributes not only to the academic understanding of the current phenomena but also to the neighborhood governance praxis.

1.6 The Caveats

Studying ongoing processes carries particular risks for the researcher: they can change rapidly in unpredictable directions. This is also the case with the virtualization of the urban environment. This process is still underway and it is difficult to build up a reliable body of knowledge: despite the visible increase in the role of ICT in the recent urban development, it is not clear whether online conversion of services and digitalization of decision-making processes will be a permanent trend or if it will turn out to be just a temporary fashion.

This uncertainty should not discourage us from drawing conclusions and providing solutions for urban governance however. Anticipatory character of urban planning requires particular prerequisites, based on observation of the current trends.

That is the reason for undertaking bold and controversial research, even if it brings only superficial results at first.

Anyway, the interpretations, assessments, and concepts presented in this book should be treated with appropriate caution. As a reaction to the ongoing processes, they may miss some forthcoming pitfalls of urban virtualization and as a result may lead to irrelevant practices in neighborhood governance. The idea of multichannel neighborhood governance will have to be tested and verified in policy practice and as a result can be confirmed or neglected.

1.7 The Foundations

The aim of this book is to lay the foundations for a revised neighborhood governance. Like all foundations, it is embedded in a kind of "ground"—a particular body of knowledge from which a new conceptual construction begins. In the case of neighborhood governance, this ground includes urban planning theory, community development issues, participatory governance, services provision, quality of life, administrative cultures and practices, smart city principles, and perhaps others. Some of these are universal, while others have a local scope, but all of them form the environment for governance performance.

The proper foundations start with footing layer which is located below the ground. When neighborhood governance is considered, the footing is formed by traditional approaches to local urban governance (Fig. 1.2). It involves reviewing and synthetizing all the relevant concepts and practices that have shaped the neighborhoods until recently.

The second layer is the foundation wall. It is situated on the footing and partly goes beyond the ground level. In the context of neighborhood governance, it includes its recent development connected to ICT, often presented as e-planning.

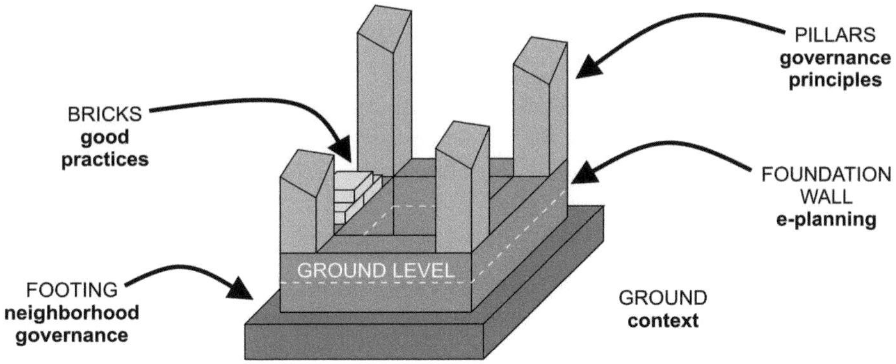

Fig. 1.2 Layers of foundations for multichannel neighborhood governance. (Source: Author's own research)

This concept is based on the traditional urban policymaking enhanced with the tools rooted in the virtual world.

The pillars supporting the future construction are located on the foundation wall. In the allegorical image drawn here, they represent the revised principles of urban governance. The way multichannel neighborhood governance is performed is embedded in both the footing (traditional urban governance) and in the foundation wall (e-planning framework).

The general methodological approach adopted in this book is a critical revision of the state-of-the-art. Such revision obviously requires a thorough review and frequent referring to various sources, including scientific literature, expert reports, and policy documents. This well-grounded and commonly appreciated research practice may seem to be generic, but when properly ordered and systematically implemented throughout the text, it becomes a solid basis for creative reinterpretation of the existing knowledge and for developing novel concepts and solutions.

It is noteworthy that foundations are only a starting point for the future building. The projected construction should be completed with bricks of particular policy practices. Such practices should reflect local organizational culture, be resistant to various external conditions, and most of all address the needs of citizens. Urban policy practices are laid just like bricks in the wall: each of them is based on the previous experiences and offers new added values, and all of them lay on the foundation wall.

The concept of multichannel neighborhood governance presented in this book paves the way for novel urban policy solutions, but it does not determine their final shape. Good neighborhood governance has always followed the evolution of human needs and lifestyles, looking far into the future but never losing the local perspective and political capacity. Its main characteristic is responsiveness—the ability to adapt to the changing civilizational contexts. Without this, the multichannel approach will not bring any visible improvement and will remain an ad hoc reaction to some temporary phenomena.

1.8 Structure of the Book and Synopsis of the Chapters

The book is organized in a classical way. The introduction is followed by setting the general conceptual framework for further research, and then selected findings are presented. Finally, the conclusions are demonstrated, referring to the research questions outlined in the beginning.

The analytical synopsis of the contents of each chapter is presented below.

1. Introduction. The first chapter presents the origins of the multichannel governance concept, research questions, objectives of the book, and the range of the studies presented. It also outlines the content of particular chapters.
2. Laying the foundations. This chapter recalls the core paradigms of urban governance and guides the reader through the concept of the neighborhood, local

policy approaches, and the role of ICT in shaping current urban environment. It also shows the links between neighborhood planning and urban governance.
3. Multichannel objects of governance. In this chapter, the services sector is presented taking various perspectives (geographical, social, political). Its main argument is that the shift from traditional services to online services in the recent years has a considerable impact on urban territories, including neighborhoods.
4. Governance as a multichannel process. This chapter describes the recent changes in the neighborhood governance patterns. Virtualization of various spheres of urban development bears particular consequences to the decision-making processes.
5. Conclusions. The last chapter presents the gaps in the current governance paradigms; it addresses the research questions and concludes on the role of ICT tools in both the form of neighborhood management and its substance. The concept of multichannel neighborhood governance is introduced. The chapter also points to the contribution of the book to the state-of-the-art and proposes further research directions.

Throughout the text, selected examples are presented in dedicated boxes to aid understanding. Each box contains a general description of a particular case, followed by a table or figure visualizing the problem. The text is also enriched with additional content presented in sections entitled "Practitioner's Perspective." These aim to illustrate selected issues with the practical experience of various people involved (directly or indirectly) in the urban decision-making process.

Chapter 2
Laying the Foundations: Paradigms of Neighborhood Governance Versus Current Trends in Urban Development

Abstract In the search for novel solutions for neighborhood governance, it is necessary to review the to-date paradigms and concepts first. This section starts with an overview of various manifestos in urban governance, followed by several definitions of neighborhood and different approaches to local policy. It also shows the links between neighborhood planning and urban governance. Then selected aspects of virtualization of urban environment are presented as a challenge that needs to be addressed with relevant governance solutions. The section concludes with a synthesis of the to-date urban policy paradigms and a list of issues demanding to be reviewed among the neighborhood governance priorities.

Keywords Urban governance · Paradigms · Neighborhood · Localism · Livability · Neighborhood planning · Neighborhood governance · Resilience · e-planning

2.1 Urban Governance Paradigms Revisited

As it noted above, the main ambition of this book is to verify the urban governance paradigms. By a "paradigm" I mean "a distinct set of concepts or patterns of thought, including theories, research methods, postulates, and standards for what constitutes legitimate contributions to a field" as defined by Wikipedia[1] (2020). Creating a paradigm requires a closed system that accepts changes: paradigms can only apply to a system that is not in its final stage (Kuhn 1970). This applies well to urban governance, where there are no definitive solutions. Urban development is characterized by equifinality—a given end state can be achieved by many potential means. And neighborhood governance is a way of managing this never-ending process of change.

[1] Wikipedia, an online encyclopedia presenting user-created content (opposed to expert-created content published in traditional dictionaries), is rarely used as a source of reliable academic knowledge. However, its definitions offer a good basis for effective communication between researchers and nonacademics. Hence, I decided to use this source for building the foundations of this book.

Ł. Damurski, *The Foundations of Multi-channel Neighborhood Governance*, Local and Urban Governance, https://doi.org/10.1007/978-3-031-75124-0_2

Paradigms of urban governance and planning have various origins: some of them are derived from stakeholders' reaction to particular urban developments and trends; others come from complex research performed by experts using a wide range of scientific methods; finally some are based on ambitious visions of desirable urban futures proclaimed by charismatic individuals or groups. If one wants to revise the paradigms, he or she should follow the same patterns: track the current governance practices, review various research results, and eventually collect and synthesize futuristic postulates and ideas. This is what I do in this section.

Until the nineteenth century, urban development had been guided by local, regional, and national monarchs, designed by architects, and performed by constructors. The pace and scale of development in that time did not require comprehensive coordination. This is one of the reasons why the first paradigms of urban governance emerged in the industrial epoch when urban development became faster, more complex, and less controllable.

The core paradigms relevant for neighborhood governance may be found in various "manifestos" proclaimed from time to time by urban planners and governance practitioners in the twentieth and twenty-first centuries. Referring to such manifestos has particular assets: (i) they provide a diagnosis of contemporary urban issues; (ii) they reflect the needs, aspirations, and priorities of particular generation of urban communities, giving a valuable insight into what were the perceived deficits or threats that needed to be tackled; (iii) they are prepared collectively by groups of thinkers (not by individual "urban prophets") based on a thorough reflection on the current condition of urban areas; (iv) they provide innovative ideas for urban governance that transcend the current thinking frameworks. Even the titles of those documents express the evolution of urban thinking from technocratic "rules of urban planning" toward strategic "visions of cities and regions" (Mironowicz and Majda 2017).

A brief overview of the most important manifestos in urban planning enables listing some core threads that are relatively common to most of those documents. This overview starts with three basic questions that reflect the scope of this book: (1) how does each manifesto situate the neighborhood in the urban structure? (2) what does each manifesto say about urban services and citizen participation? (3) how does it refer to technologies, including ICT and the virtualization of the urban environment? In what follows, the questions have been answered by referring to specific documents, thus demonstrating not only specific ideas but also the progress in urban governance paradigms.

Another source of urban governance paradigms is the current documents of urban policy. Opposed to the manifestos described above, they provide particular instructions that need to be followed by policy practitioners and public authorities at various levels. On one hand, they are some kind of adaptation and implementation of the current governance paradigms, being their result and emanation, but on the other hand, they develop some practical aspects to the futuristic visions presented in "urban manifestos." Due to their imperative character, various reports, handbooks, guides, and acts impose relevant policy practices, and as such, they complement the general principles for urban governance with additional, pragmatic values. In this section, selected urban policy documents will be referred as a complementary material in the search for governance paradigms.

2.2 Selective Review of Urban Manifestos

Below selected results of the review of urban manifestos are presented, answering the questions on the role of the neighborhood, urban services, citizen participation, and ICT. Particular statements have been rephrased for better readability, followed by reference to particular document. All the relevant paradigms are summarized and synthesized at the end of this section.

1. Urban functional structures, neighborhoods, and environment.

   - "The initial nucleus of urbanism is a cell for living – a dwelling – and its insertion into a group forming a habitation unit of efficient size (…). With this dwelling unit as the starting point, relationships within the urban space will be established between habitation, work places and the facilities set aside for leisure" (Le Corbusier 1973, pp. 101–102).
   - "Housing and work areas, as well as other compatible uses, should be closely related in time and space so as to reduce the need to travel, conserve energy and reduce pollution" (ECTP 1998, p. 16).
   - "Housing and services should become increasingly accessible – their provision will be flexibly adjusted to new and emerging patterns of needs. More housing will be provided at affordable prices, in addition to educational, commercial, cultural and recreational facilities and services. These will be supported by running costs that citizens can afford, and complemented by a strong sense of community identity and security" (ECTP 2003, p. 5).
   - "Human activities must be located within the urban core and the tendency to sprawl into the hinterland, consuming rural and natural areas must be reversed" (ECTP 2013, p. 11).

2. Urban services and accessibility standards.

   - "The advantages of mixed-use areas have to be considered very seriously, against the more traditional approach of concentrating commercial and residential uses in designated zones. It is appropriate to note that, in land use terms, that mixed-use areas, already found in older cities may provide diversity, coupled with increased social and economic activity. In turn this enhances the overall quality of life" (ECTP 1998, p. 6).
   - "In terms of land use (…) the most radical effect of technological development may be to eliminate the need for large scale offices and industrial structures, thus reducing the demands for space in cities. In turn, this may facilitate the process towards more mixed development and greater social interaction" (ECTP, 1998, p. 4).

- "Planning should examine the possibilities for decentralization of activities, taking account of new technology, envisaging the development of a polycentric, multi-faceted city (…). The disaggregation of activities, both in time and place, should be encouraged" (ECTP 1998, p. 13).
- "In the connected city and its regional hinterland, new technologies will be applied creatively to provide a variety of systems of transportation of persons and materials, and of information flows. At the local scale, technology and traffic management will be deployed to secure a decrease in the reliance on private vehicles" (ECTP 2003, p. 5).
- "In the connected city, the essential functions of city centers and other key nodes will be maintained and improved; communications and transportation networks will serve these efficiently, without allowing the latter to sap their vitality" (ECTP 2003, p. 9).
- "Ease of movement and access will be a critical element of city living, together with greater choice in the mode of transport, for example, through better information or virtual access enabled by new technologies. There is also a need to minimize travel distances for basic goods and services through efficient, pleasant, sustainable, and economical connections between places" (ECTP 2013, p. 14).

3. Technology effects and social change.

   - "To accomplish this great task (urban planning and management – added by Ł.D.), it is essential to utilize the resources of modern techniques, which, through the collaboration of specialists, will support the art of building with all the dependability that science can provide, and enrich it with the inventions and resources of the age" (Le Corbusier 1973, p. 102).
   - "The revolution in information technology and electronic communications is already having a marked effect on the ways in which cities operate. Further changes may be expected to reduce the overall need to travel, to change the nature of the workplace, and to enhance the capability of citizens to obtain information quickly and to communicate effectively" (ECTP 1998, p. 4).
   - "In overall terms, the information revolution will probably have a positive effect on the future development of the city. Nevertheless, there will be a need to guard against the possible negative consequences, including social isolation and divisions between the information-rich and the information-poor" (ECTP 1998, p. 4).
   - "The universal development of information technology has great implications for social change. Nevertheless, personal contact will remain important, for which electronic communications are not a substitute. Rather, information technology increases the possibilities for communication and the diversity of experience (…). The information network should play a key role in providing a framework for human contact, which is a basic component of cultural identity and social cohesion" (ECTP 1998, p. 13).

- "Planning should encourage the optimum use of information technology, with equitable access, so as to obtain the maximum benefits for the citizen" (ECTP 1998, p. 13).
- "The spatial organization of the connected city will include a full integration of transportation and town planning policies. They will be complemented by more imaginative urban design and easier access to information, thus minimizing the need for unnecessary travel" (ECTP 2003, p. 5).
- "The development and growth of the knowledge-based economy has changed dramatically the driving forces of urban development in Europe. Advanced 'service producers' are becoming the leading activity in cities, while universal access to networked computer-based facilities allows for working at home, e-commerce and e-business" (ECTP 2003, p. 13).
- "Working on the web from home as well as e-commerce and e-business may result in less need for physical urban facilities. On the other hand, these processes may generate much more traffic for goods movement and delivery affecting already congested city centres" (ECTP 2003, p. 13).

4. Citizen participation.

- "Future European cities will be used not only by resident citizens, but also by other consumers of their facilities and services on a permanent or a temporary basis (commuters and visitors) (…). Sufficient time must be built into the decision-making processes relating to spatial planning and development, so that social links can be established, and positive interactions facilitated" (ECTP 2003, pp. 3–4).
- "The democratic processes may be enhanced by providing information to those who traditionally did not have access to it. Potentially, it can enable the citizen to become involved in the management of the city, provided that there is equitable access to resources (…). New technology can also provide opportunities to focus on themes or issues of common interest for all citizens—whether city-wide, or the neighborhood where they live" (ECTP 1998, p. 13).
- "It is widely recognized that planning is not solely concerned with plan preparation. It is also part of a political process aiming to balance all relevant interests—public and private—so as to solve conflicting demands on space and development programs" (ECTP 2003, p. 17).
- "Planning enables communities to formulate strategic visions to achieve their future aspirations. Such visions have the potential to have significant and direct impacts creating great places and delivering a more sustainable future for communities throughout Europe" (ECTP 2013, p. 6).
- "Spatial integration in appraising tasks and setting policy requires an integrative pro-active and participatory approach to design, communication and monitoring. The approach required depends on who is involved, the stage of the planning process, and the planning or political context. In some situations an informal and more experimental approach may be desirable or necessary" (ECTP, 2013, p. 25).

5. Manifestos summary.

   The overview of selected aspects of urban manifestos presented above enables formulating several key principles in managing urban development. These are as follows:

   - Neighborhood is a basic unit of each urban structure.
   - Housing areas should be closely related in time and space with workplaces; mixed-use areas should be promoted.
   - Accessibility of services should be increased by using new technologies and new mobility patterns to minimize the need for traveling and provide desired quality of life.
   - A strong sense of community is important; social isolation should be mitigated.
   - Urban sprawl tendencies should be reversed.
   - Polycentricity of various activities and functions should be encouraged; urban nodes should be improved.
   - Democratic standards in decision-making should be enforced, balancing all relevant interests and providing citizen participation, communication, and monitoring.
   - New technologies should support better planning and management practices.

The role of new technologies is clearly acknowledged in the analyzed documents. ICT is expected to alter the traditional functional structures of urban areas (e.g., by reducing the need for large scale offices) and influence urban communities (with easier human interaction but lower social cohesion) and policy practices (offering direct access to comprehensive data and new management strategies). Such visible presence of the technological aspect of urban governance proves that the manifestos have fulfilled their task: they provided a conceptual foreground for future adjustment of policy practices.

However, the documents presented in this section have lacked a holistic approach to the contemporary neighborhood. Therefore, the aim of this book is to start a discussion on the vital interrelationship of online and offline urban environments in the neighborhood context. To enable such a discussion, the general, citywide scope of urban paradigms needs to be narrowed and adapted to the local scale. And such an adaptation requires a solid conceptual basis, including definitions of the key terms involved. The following sections present some reflections on the concept of neighborhood, followed by a brief overview of selected approaches to neighborhood governance.

Box 2.1: "Complete Neighborhood" Program in Wrocław

Since 2019, Wrocław municipality has been developing an novel policy program titled "Osiedla kompletne" ("Complete neighborhood") which aims at supporting local communities in improving their neighborhoods. It provides a set of recommendations for managing local spaces, allocating various functions to particular elements of the neighborhood. For example, a square (red dot on the map below) plays several roles: it stimulates the social life of the local community, offers everyday services, strengthens urban layout, fosters biodiversity, and accumulates water. All those complementary functions articulate the philosophy of "neighborhood completeness" (Biuro Rozwoju Wrocławia, 2019). The program follows several universal principles of urban governance, including citizen participation, resilience, sustainability, and short walking distances. However, it does not address the issue of conversion and hybridization of services.

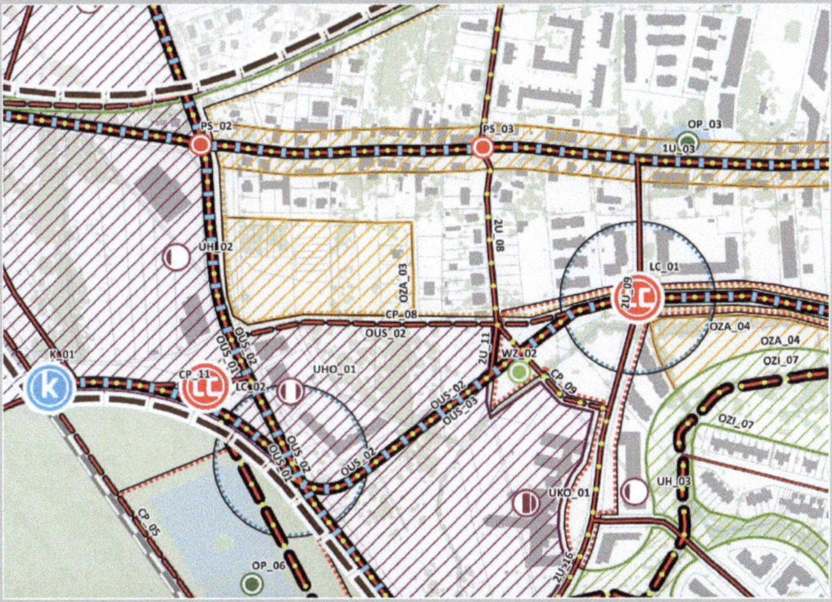

Part of the map of recommendations for Maślice Małe complete neighborhood (Biuro Rozwoju Wrocławia, 2019). (Permission for reprint obtained from the copyright owner on 18/09/2023. Origin: Author's own research)

2.3 In the Search for Neighborhood Definition

In tracing the historical notion of the "neighborhood," we can turn to the ancient urban structures described by Mumford (1961). The excavations in Mesopotamian city of Ur proved that the city consisted of a series of more or less coherent subareas in which smaller shrines and temples were located. Such subareas were also familiar to Greek city planners who divided the city into definite "neighborhoods" which were deliberately fabricated physical entities.

However, the neighborhood of Greek planned cities and later of Roman cities appears to be unlike what we refer to as neighborhood today. They appear to primarily have been physical structures, designed to provide a certain orderliness to the overall urban fabric. The social and cultural elements that seem to be embodied in the modern notion of neighborhood probably have been absent in the ancient times (Peterman 1999).

Current research suggests that the idea of neighborhood is of more recent origin and that its roots can be found in the suburbs of the Western cities. Urban social reformers at the end of the nineteenth century and in the beginning of the twentieth century saw suburbanization as a vital force and viewed it as an antidote for urban social ills.

Practitioner's Perspective
What are the neighborhood characteristics?

Neighborhood (…) includes senior housing, health facilities, affordable housing, and so on. Walking distance is very important in neighborhoods.

Donald W. Birch, Executive Vice President/Chief Operating Officer Leggat McCall Properties LLC [interview conducted via MS Teams 9/09/2021 11:00 am-12:30 am; origin: Fulbright Senior Award scholarship 2021]

When the Regional Planning Association of America was formed in 1923, it adopted the Ebenezer Howard utopian garden city model for creating a sense of community in residential neighborhoods. Shortly after the RPAA was formed, Perry (1929/1998) began promoting the concept of "neighborhood unit" as a fractional urban unit that would be self-sufficient yet related to the whole.

Later, during the depression of the 1930s, planning profession in the USA fully embraced the notion of neighborhood unit, supported by the newly organized University of Chicago school of sociology, which saw it as a promising mechanism for maintaining urban stability (Peterman 1999). Thus, although the concept of the neighborhood as a "village" within a larger urban agglomeration may have ancient historical antecedents, its conceptual roots are more recent and are associated with the creation of middle-class "residential heaven."

Perry's ideal neighborhood unit was to be relatively self-contained, having its own residences, shops, churches, and recreational facilities. The rules of a quarter- to a half-mile of radius, about 5000 people, and the central location of a core facility (elementary school) were the key organizing factors of the area. Major arterial roads did not pass through residential areas but provided the boundaries of the neighborhood so that children did not have to cross traffic streets on their way to school (Wojtowicz 1998) (Park and Rogers 2015). In effect, the residential portions of cities were to be assembled from such modules.

Despite the changes in urban structures introduced in the twentieth and early twenty-first century, neighborhood unit theory remains the main reference for urban research and governance all over the world. The values of limited population size and urban layout determined by pedestrian accessibility to the community center are still popular among urban governance theoreticians and practitioners as crucial components of a livable neighborhood.

In the search for a more complete explanation of the term, some examples of neighborhood definitions are presented below, pointing to different physical, social, and functional aspects (selection according to (Park and Rogers 2015)):

- An "important organ of urban life," in which people are connected, interrelated and interdependent like all living organisms (Mumford 1954).
- A combination of geographical boundaries, ethnic or cultural characteristics of the inhabitants, psychological unity, or concentrated use of an area's facilities (Keller 1968).
- A small urban area where residents are influenced by socioeconomic effects and services within (Goodman 1977).
- A collection of spatial attributes associated with clusters of dwellings, sometimes in conjunction with other land uses (Galster et al. 2001).
- A particular form of social reproduction in which human activities, including daily living, social interaction, and political and economic engagement, take place (Martin 2003), (Park and Rogers 2015).

Hence, conceptually, neighborhoods are both geographical and social units: they involve some kind of interplay between place and people (cf (Ohmer et al. 2019)). Taking a geographical perspective, a neighborhood needs to be large enough to support services, while using a social perspective—it needs to remain small enough for residents to maintain a shared identity (Weiss et al. 2007). Moreover, as a subarea of a town or a city, it is distinguishable from other neighborhoods due to its specific physical and social characteristics (Rohe 2009). In other words, a neighborhood is a unique collection of people who share particular services in spatial proximity and some level of social cohesion in a geographically bounded place.

Attachment to a particular place is an elementary criterion defining a neighborhood and differentiates it from other similar terms. For example, the word "community" though apparently synonymous to neighborhood refers to a group of people with a unity of values, beliefs, interests, and culture, but without a clear geographical reference (Chaskin 1997), (Keller 1968). It is more attributed with circumstantial (socioeconomic status and life cycle) and functional (shared services and goods) aspects than social (a network of relationships) and cultural (religion or ethnicity) characteristics (Park and Rogers 2015). As some authors note, neighborhoods are communities with a tangible spatial dimension and as such can be useful for governance purposes (Wellman and Leighton 1979; Forrest and Kearns 2001; Mullan et al. 2004).

The concept of neighborhood can be considered at multiple scales, depending on the size, spatial structure, and services of a particular residential area. A hierarchy of neighborhoods developed by various authors (Chaskin 1998; Suttles 1972) based on their physical conditions demonstrates that the size of neighborhoods can vary from a few households to large communities of thousands of residents. Park and Rogers (2015), referring to the works of American Planning Association, presented particular requirements for each level of the neighborhood hierarchy:

- Face blocks are small neighborhoods of houses on two sides of a street. The physical proximity allows for strong interpersonal ties, as residents tend to have some degree of personal relationship, or at least facial recognition.
- Residential neighborhoods consist of several face blocks and usually share certain public and commercial services (including primary school, nursery, or community center), green areas, public spaces, and transport hubs. They are relatively homogeneous in terms of the socioeconomic characteristics of their residents (demography, ethnicity, income, education). Local identity in residential neighborhoods is represented by specific names and urban landscapes. Direct participation of residents in planning processes is possible due to the relatively small number of residents, ranging from 500 to 5000 people on 15–500 acres of land.
- Institutional neighborhoods are constellations of residential neighborhoods plus other categories of land use. At this level, official neighborhood authorities are established and exercise their powers within a clearly defined spatial boundary. The institutional neighborhood must be large enough to provide certain services (shopping centers, schools, health centers, hospitals, government offices, financial institutions, and recreational facilities). The size of such neighborhood is usually between 5000 and 10,000 residents and about 1000 acres of land (Park and Rogers 2015).

Neighborhood can be differently operationalized for each analysis. Empirical research usually defines neighborhoods in such a way which maximizes data availability and increases the likelihood of achieving meaningful outcomes. Feasibility of collecting data on selected neighborhoods, including sampling, ease of observation, or data availability are crucial factors in specifying the range of a neighborhood (Ohmer et al. 2019; Park and Rogers 2015). Hence, there are no uniform criteria for all, and every study can adopt a unique definition and delimitation of neighborhood. Moreover, as we are entering the era of transient societies, where local communities are increasingly mobile, urbanized, and international, the neighborhood also becomes a transient concept (cf (The Young Foundation 2010; Power 2004)). The growing role of online communication results in less in-person interaction, lower social control, and finally reduced place attachment. Accordingly, the concept of neighborhood as a core component of urban structure may be diminished and replaced with more relational ideas, limited to merely functional, aspatial aspects. To make it more clear, an overview of approaches to neighborhood boundaries will be presented below.

2.4 Neighborhood Boundaries

Delineating a geographically specified neighborhood is not an easy task. Although we may live in a location which we may refer to as our neighborhood, the boundaries of this self-defined place are not likely to be exact. Furthermore, what we consider to be our neighborhood may be quite different at different times and for different purposes (Peterman, 1999). The meaning and the range of neighborhood is tied to some attributes valued by the residents in a defined territory. Where a group of people values the same attributes of the same area, we can call that area a neighborhood (Galster 2001).

Neighborhoods are often determined by immaterial elements such as population, area, or functional profile. However, in the classical approach to urbanism, it is the physical boundary that is a critical condition in defining neighborhoods: it protects encroachment from outside; it unites residents residing in and creates places for transactions with surrounding functions (Alexander et al. 1977a, b). Natural boundaries (e.g., farmland, forest, water, and green ways) and man-made barriers (e.g., railroads, major roads, parks) are frequently mentioned in the literature as the material borderlines for local urban communities.

But the vast majority of studies use administratively defined units as proxies for neighborhoods, even though administrative boundaries are established for statistical or programmatic purposes and do not necessarily coincide with individual's experience with regard to their neighborhoods (Ohmer et al. 2019). The main asset of such approach is that administrative division is usually followed by some kind of political power and enables establishing local decision-making bodies. However, there still is a need to develop also some other methods of delineating neighborhood borders.

For example, resident perceptions may be a good alternative, representing the areas that are relevant to people's everyday lives. In this approach, it is necessary to reach a consensus on the residents' perspectives on neighborhood boundaries as each member of a community has a different notion of the place (Ohmer et al. 2019).

Another approach to determining the aerial unit for community measurement is to map the areas in which individuals spend time. This method recognizes that as residents move through their surroundings, they carve their own activity space that does not necessarily match their residential neighborhood alone or comport with predetermined geographical boundaries (Ohmer et al. 2019). Jones (1990) describes this method in the following way: "residents can be asked a (…) set of questions about how they use the neighborhood: where they walk or do not walk, where they drive or do not drive, where they shop for various items, what park they use, and so on. Small businesses can be asked about the area from which they draw their customers (…). When all those pieces of information are plotted on maps, one starts to get some sense of the common area that means something to people" (Jones 1990, p. 9). Once the unique activity space is determined for individuals, it is possible to go back to the origins and to measure functional, physical, and other characteristics of the given area.

Also, remembering that neighborhoods are fundamentally spatial and therefore require more than just social and behavioral characteristics, some scholars delimitate neighborhood boundaries using the quantity and diversity of functions and land uses (Rohe 2009). The resulting functional clusters reflect relatively well the range of urban activities performed in particular areas and enable a meaningful division of the city into smaller units. Simultaneously, they also reveal a significant discrepancy between the traditional administrative partitions and the city's factual organization. Functional clusters present "more homogeneous distribution and degree of specialization in terms of the quantity and diversity of urban activities, when compared to the administrative neighborhoods" (Martí et al. 2022, p. 89).

On the intersection of physical and social characteristics of the neighborhood, the concept of localism is situated. In the following section, this term will be analyzed in detail, offering an important input into the understanding of the neighborhood.

2.5 Localism and Place Attachment

People can be aware of various geographical units, but neighborhood is a particularly important one, because it contains the key social and person-place transactions (Taylor et al. 1984). Neighborhoods are the most recognizable units of local identity, offering the basis for an emotional attachment of residents to particular places. Among many various aspects of this multifaceted relationship between space and community, four seem to be crucial for a proper consideration of contemporary neighborhood values.

First, there is a dearth of empirical research examining the links between the sense of community and the land use of particular area. The perceived quality of neighborhood services, public spaces, and various facilities is significantly and positively associated with the sense of community (Francis et al. 2012).

Second, place attachment involves positively experienced bonds, sometimes unconscious, that develop over time from behavioral ties between people. These ties provide a framework for both individual and communal aspects of identity (Cattell et al. 2008). In addition, neighborhood citizens tend to have a relatively homogeneous socioeconomic status and a sense of fraternity that contributes to social embeddedness in the local community. They also have the political capacity to protect the interests of the neighborhood and to explain what they need (Park and Rogers 2015).

Third, a research conducted by Lewicka (2012) in various settlement contexts shows that people attached to particular place of residence are rooted, live longer in a given place, have closer relationships with neighbors, trust other people, willingly engage in various community activities, are interested in the history of their place of residence, and generally present higher levels of life satisfaction (Lewicka 2012).

Practitioner's Perspective
Place that does not exist offline

- *Hello, we have found "Karmelska Beach" on Google Maps. Could you please tell us if it is nice and how to get there?*
- *I don't know any "Karmelska Beach" here. Google Maps sometimes gives names to some places which are not recognized by the locals.*

A dialog with Elizabeta, a landlord in one of the Croatian tourist resorts [Croatia 2023; origin: author's own research]

Fourth, the so-called third spaces like cafés and pubs where people can meet old friends, make new acquaintances, or discuss the important issues of the day are vital settings for public life. Involvement in informal social activity has important psychological, social, and political implications; therefore, such ordinary spaces are a significant resource for both individuals and communities (The Young Foundation 2010). By enabling social interaction, they can provide opportunities for exchange of goods and ideas, relief from daily routines, and sustenance for people's sense of community and can raise people's spirits. They also possess subjective meanings that accumulate over time and can contribute to neighborhood governance (cf (Cattell et al. 2008)).

The four aspects—the interrelatedness of land use and the sense of community, the social ties within the community, the rootedness resulting in mutual trust, and socializing opportunities which foster citizen involvement—form some kind of basis for further discussion of various developments related to place attachment within a neighborhood.

Taking a historical perspective, we need to note that localism has functioned in human mentality for a very long time. Dividing the world into "my own" and "foreign" localism was associated with the original delineation of spatial boundaries, the creation of the first mental maps, and it existed long before administrative boundaries were set. Local communities ruled a specific territory and built a relatively closed spatial, social, and political system. Somewhere above them, wars were fought, monarchies collapsed and emerged, and political macrostructures were formed (Sowa 2000).

The decline of European localism began in the nineteenth century and culminated in the mid-twentieth century (Castells 1999). This was mainly due to processes of industrialization, urbanization, and suburbanization, as well as various political factors. Massive migrations to work in industry broke the continuity of local social structures, suburbanization created new communities living in "places without history" (Leboreiro Amaro 2014), and the emergence of welfare state entailed a growing role of the state in economic and social life (Sowa 2000). Culture is also an important element of localism, including local traditions, local artistic activities, customs, and habits. Therefore, the weakening of local communities was also associated with the growing role of mass media—the press, radio, and from the mid-twentieth century also television—which led to the erosion of social norms (Sowa 2000). None of this was conductive to localism and the self-government of local communities.

The subsequent comeback and rise of localism have been linked paradoxically to increasing globalization. As power flows upward and outward from national governments to complex networks of quasi-state agencies, a "democratic deficit" appears, resulting in a disengagement of the public with politics, intensified by cultural and political fragmentation of contemporary societies. In this context, the local becomes the space where publics can be reengaged and experiments in governance can be carried out (Brownill 2017).

Practitioner's Perspective
What does "local" mean?

It is important where the product is coming from. When the product is coming from the store, not from the distribution center, it is a "local" product. It can be picked up by the customer, or it can be picked up by a third-party service that will take it to the customer's home, but it is local. There is some kind of anchor in the neighborhood. Local product in that context means that the inventory is held locally (not necessarily that the product is produced locally).

Antonio Moreno, Sicupira Family Associate Professor of Business Administration, Harvard Business School; Unit: Technology and Operations Management [interview conducted via Zoom 11/04/2021; reviewed and authorized by the interviewee on 14th of November 2021; origin: Fulbright Senior Award scholarship 2021]

According to F. Fukuyama (2000), one of the most unusual, surprising phenomena of our time is the revival of social self-organization and the renaissance of localism. This process has been conditioned primarily by the crisis of the state and more precisely of the policy excessively interfering in local economic and social matters. The inefficiency of administrative structures has contributed to the development of the informal economy, which is an important factor in the revitalization of local systems (cf. (Leśniak-Moczuk 2007)).

Box 2.2: Mała Czarna—the Local Coffee Roasters

A network of small cafés in Wrocław uses the word "local" to promote its outlets. The network is developed on the growing trend of socializing around cafés and restaurants. Although coffee is not grown locally in Wrocław (and in general not in Europe), it can be locally roasted and served. This approach is supposed to carry positive connotations with a particular place and with a particular community. Apart from this marketing strategy results, Mała Czarna really contributes to the neighborhood services by offering attractive meeting places for the residents.

One of the Mała Czarna outlets in Wrocław. (Source: Picture taken by the author in 2023. Origin: Author's own research)

In Central and Eastern Europe, the development of localism was additionally associated with the decline of socialism and the following political transformation to free market and democracy. Those processes firstly led to the growth in personal freedoms but then were followed by the empowerment of local communities (Sowa 2000). Ideals of power decentralization have been particularly appreciated in Poland, where after the period of political transition of the 1990s and 2000s and solving the most pressing structural issues (such as downtown revitalization, deindustrialization of urban economy, or building nationwide network of highways), the focus and attention of local authorities shifted toward the neighborhood level (Damurski et al. 2020). We are witnessing the revival of localisms and regionalisms, the revival of territorial government movements (Jałowiecki and Szczepański 2002).

Localism is becoming not only a fashion, not only an opportunity for development, but a kind of worldview that is practiced by more and more citizens, entrepreneurs, managers, and politicians (Jałowiecki, 1989). Many municipal policies are currently created on the basis of this concept: there are local brands, local products, support groups, activities, and taxes. Today, localism is a kind of buzzword in the global discourse on urban governance.

Place attachment is anything but uniform, stable, or static: It can be a resource for the local community, but it is also fragile and can be reduced or destroyed. While it is possible for place attachment to be enhanced through neighborhood governance which provides a counternarrative to NIMBYism and posits neighborhood as a field of care (Bradley 2017), it is also possible that the opposite may occur when the localism feelings can be so strong that rational deliberation can become difficult (Banfield 2020). For this reason, a careful approach to the relationship between localism and local governance is needed.

2.6 The Role of the Name

One of the key aspects of place attachment is related to naming. Names are integral to the geographical construction of place (Tuan 1991). Even though modern societies are less reliant on local geography for physical survival than their predecessors, they still identify with the place where they live and give it a unique name (Taylor et al. 1984).

Urban identities are complex and fragmented along many lines, however. Identification of space by the users varies in scale, ranging from individual houses to whole neighborhoods and larger public areas. Indeed, it is difficult to grasp the entire city as a whole—it is not mentally "digestible" for the residents, and does not fit their perception (Bagiński 1998). As a result, conveying a single citywide identity becomes nearly impossible. Instead, we see the growing importance of finer distinctions between different parts of the city, reflected by names that distinguish them from adjacent areas (Reitman 2004).

Application of a name to a locale is also important for building person-place bonds. "In functional terms, it is not only the direct experience of the terrain which assists the individual in building up his spatial world; language crystallizes this knowledge through the customary use of place names" (Hallowell 1955, p. 193). Taking language seriously leads to an observation that the "quality" of place is more than just esthetic but also has a moral dimension (Tuan 1991).

Hence, the label of a place is filled with various contents and conveys these contents to inhabitants and to people from the outside. As such it offers a formal basis for both internal and external functional relations of the community (Helleland et al. 2012; Jordan 2020). Naming has also the power of defining the identity of a place. To change the name is to change, subtly and inexplicably, the place itself (Tuan 1991).

To sum up, there are several roles played by place names: they emphasize spatial characteristics that are important for a community and support the "personality" of a place, they mark particular territory, they structure the space and constitute users' mental maps, they enable the process of placemaking, and they support emotional ties between humans and space and create the space-related identity both of individuals and communities (Jordan 2020; Tuan 1991). As Tuan noted, "the feel of a place is registered in one's muscles and bones" (Tuan 1977, p. 184).

Additionally, a neighborhood name empowers residents and helps them to draw attention to their area. It provides an umbrella under which local stakeholders can represent their interests to the decision-makers, allowing them to establish political legitimacy and to have a say. In this context, naming may also be a marketing tool. In the global competition for residents, businesses, and investment, branding becomes a powerful strategy for creating a "place." A properly named neighborhood may attract the attention of the desired target groups, allowing its community to present the local identity and to project a collective vision for the future (Reitman 2004).

Neighborhood name formulates a linguistic framework for performing local community tasks and is an essential condition for place attachment of the residents. Areas without a specific name are perceived just as parts of other entities. If a neighborhood is subject to localism—as I have argued in the previous section—it has to have a specific name. This observation brings us to the reflection upon neighborhood cohesion, as a concept embracing various territorial and social values of a residential area.

2.7 Neighborhood Cohesion

The concept of neighborhood cohesion was originally an amalgamation of several approaches within social and psychological sciences. The term developed initially by Buckner (1988) was a synthesis of psychological sense of community, attraction-to-neighborhood, and social interaction within a given residential area. Residents living within a neighborhood were expected to present some degree of cohesion.

Taking this perspective, we may note a specific role played by both family and neighborhood environment in upbringing children. Neighborhood cohesion may be observed in the relationships among parents, expressed in mutual trust, helping each other in an emergency, watching the children, and having safe places for children in the neighborhood (An and Western 2019) (see also (Zhu et al. 2014)).

Buckner's (1988) way of measuring neighborhood cohesion enabled bringing out differences between neighborhoods. These resulted from specific street structure or the location of central places within a neighborhood. However, those morphological features of a neighborhood were not explicitly quantified in the Buckner's method, offering only a descriptive reference for comparative sociopsychological analysis.

An alternative approach to neighborhood cohesion adopts a territorial perspective. "Territorial cohesion" plays a crucial role in the European Union as an overwhelming framework for comparative research and for building development strategies. In particular, the Third Cohesion Report (European Commission 2004) stated that people should not be disadvantaged by wherever they happen to live or work in the EU. Following this definition, territorial cohesion policy strives for a just distribution of opportunities in space, in order to provide room for growth, competitiveness, employment, and sustainable development (Nosek 2017; European Union 2011; Faludi 2007). Eventually, territorial cohesion may be viewed as a set of features of a particular area: a given territory presents a higher or lower level of cohesion which means a higher or lower number of available opportunities in economy, environment, transportation, services, etc. (ESPON 2012).

In the search for a comprehensive, joint social and territorial approach to neighborhood cohesion, the functional program, accessibility of essential services (both public and private), and relations between demand and supply should be considered as a basis for building social ties within a local community (Damurski 2022). As a result, an integrated vision of neighborhood cohesion may be proposed, including both social and spatial aspects (Table 2.1), contributing to the complementarity of the two approaches.

Such revision of the concept of neighborhood cohesion is necessary in building a holistic conceptual framework for urban governance. By involving both sociopsychological and geographical aspects, it makes the political idea of territorial cohesion more applicable at the local level (Damurski 2022). Also in this book, neighborhood cohesion with all its inherent values and nuances is regarded as one of the core terms enabling a more precise and reliable definition of a neighborhood itself. And it is an important reference for more abstract considerations related to virtualization effects.

Table 2.1 Juxtaposition of neighborhood cohesion measures in sociopsychological sciences and in geographical sciences

Social and psychological measures		Geographical measures	
Measure	Resulting neighborhood characteristics	Measure	Resulting neighborhood characteristics
Psychological sense of community	Social ties and social integration	Functional balance	Self-containment and functionality
Attraction-to-neighborhood	Territorial identification and local identity	Accessibility of services	Quality of life and walkability
Social interaction within a neighborhood	Social capital and civic values	Spatial relationships between supply and demand	Local market condition and adequacy of spatial structures

Source: (Damurski 2022, p. 4649). Origin: NCN research project 2012–2015

2.8 Neighborhood Livability

Neighborhood livability refers to the living environment, which should offer an acceptable quality of life to the residents. The word "acceptable" implies a strict link to the perception of local economic, social, and cultural conditions by the users (Arundel and Ronald 2017) which makes neighborhood livability a relative, mostly subjective concept. This impression is even strengthened by the interpretation presented by Van Neste and Sénécal (2015) who consider livability as the citizens' right to safe spaces for walking and cycling and to public transit.

In spite of those interpretive deliberations, it may be noted that there are also particular physical aspects of livability, such as technical infrastructures, public facilities, state of the environment, and land-use patterns. Availability, accessibility, and affordability of services have a positive relationship with neighborhood livability. The functional diversity of urban tissue (mixing workplaces, housing, services, business) attracts various people for various purposes, in various times of the day, creating a harmonious whole and determining the sociopsychological well-being of residents (Pandey et al. 2013; Wojnarowska 2017). Neighborhoods characterized with high density of housing accompanied with various amenities are usually perceived as more satisfactory living environments compared to sprawled suburbs (cf. (Lovejoy et al. 2010)).

Eventually the most common measure used in empirical studies to assess livability is neighborhood satisfaction, connected with safety, quietness, neighbor ties, accessibility, and attractiveness (cf (Lovejoy et al. 2010; Mouratidis, 2018)). Neighborhoods offering easy access to urban facilities, public spaces, and public transport present higher values of resident satisfaction. However, many other, more objective indicators of neighborhood livability are also available.

Practitioner's Perspective
Retailers want to go where people go

There has been a flight toward quality environments. Retailers are becoming more "discriminating" in where they go. They want to go where people want to be. And so environment and neighborhood matter a lot more than they used to (…). The creation of an immersive environment is critical and fundamental for a retailer's success: there is foot traffic there, there are people living there, there are people working there, there are people visiting there, and the retailers want to be there as well.

Patrick McMahon, Senior Vice President, Regional Development, Federal Realty Investment Trust, Somerville [interview conducted via Zoom 12/06/2021; reviewed and authorized by the interviewee on 11/19/2021; origin: Fulbright Senior Award scholarship 2021]

For example, the Livability Score developed by AAR Public Policy Institute (2021) rates the overall livability of a selected neighborhood (city, county, or state) on a scale from 0 to 100. It adopts seven measurement criteria—housing, neighborhood, transportation, environment, health, engagement, and opportunity—each of which also ranges from 0 to 100. The communities are scored by comparing them to one another, and to get 100 points, a neighborhood would have to score perfectly in each of the 7 livability categories. In real world, it is very difficult, because each location has some shortages (e.g., a transit-rich neighborhood has its benefits, but it can also drive up housing prices) (AARP Public Policy Institute 2021).

Another relevant tool for measuring neighborhood livability is the NeighborhoodScout (Location Inc. 2021). It provides a comprehensive database of hyperlocal real-estate data, using over 600 characteristics, to build a neighborhood profile for every community in the USA. These include school quality; housing costs; crime rates; income levels; the age, size, and style of homes; the density of buildings; rental areas versus owner occupied; the proportion of families with children; languages spoken; types of careers of those living in the neighborhood; economic and demographic trends; unemployment trends; and many more (Location Inc. 2021).

However, none of the above-presented approaches to neighborhood livability (be it academic or commercial) includes the virtualization effects. The undisputed advantages of availability, accessibility, and affordability of traditional neighborhood services (see the definition included in the section "Services in the neighborhood") are now challenged by digitalization (in particular conversion and hybridization of services). The current market processes question the material structures of neighborhoods and require a revision of the to-date approach to livability, i.e., comprehensive studies followed by policy answers.

2.9 Neighborhood Planning

Gradually approaching the notion of neighborhood governance, I need to get back to the concept of neighborhood unit. Perry assumed that residential portions of cities would be assembled from such modules. This assembling was the work of planners, city officials, and real-estate agencies (Bennett 2017).

For this reason, neighborhood planning from its very beginning has been an undeniably political process. It is bound into political structures and procedures, and it is all about power dynamics. And if planning as a field of political practice is about the relationship between people and place, then neighborhood planning is about the relationship between a certain community and a specifically local place, but within the context of the broader planning system (Banfield, 2020).

According to Bennett (2017), the purpose of neighborhood planning can be considered in minimalist or more expansive terms. In the minimalist sense, neighborhood planning brings together professional planners and local residents to create physical plans or organize social action in a way that responds to local specifities.

Planning that answers local needs, regulates land use, locates public facilities, and delivers social services in a manner that best serves the well-being of local communities. The expansive vision of neighborhood planning assumes that it is also a tool for activating citizens, getting them to study complicated issues and engaging them in empowered participation (Bennett 2017). As such, it resembles neighborhood governance.

One of the expected outcomes of the neighborhood planning is identification of a range of community concerns and projects that local authorities can take forward (Banfield 2020). Future risks (and benefits) can be reduced (or increased) because residents identify neighborhood threats (and assets) in the early phases of the decision-making process (Park and Rogers 2015). As a result, neighborhood planning is responsive to the local needs and effectively plays a subsidiary role to the rest of the planning system.

However, planners' work is meaningful only when it is done collaboratively with the neighborhood. Neighborhood plans get implemented only if residents, with assistance from their neighborhood planner, systematically work to get the plan's recommendations acted on by the appropriate parties. There are some side effects, such as increased citizen involvement, the development of leadership among the residents, and increase in knowledge about the neighborhood. All these can be collectively referred to as community development (Jones 1990). Therefore, neighborhood planning enables local communities to produce their own development plan for their immediate neighborhood, which grants them a degree of control over what type of development takes place, and where, in their area (Banfield 2020).

A first important shift in neighborhood planning took place in the USA when the Economic Opportunity Act was introduced in 1964, followed by the rapid formation of "community action agencies" (CAAs) across the country. Until then, the relationship between those elites and the residents was "paternalistic": there was no expectation of dialog between local knowledge and trained expertise. Since the 1960s, CAAs recruited neighborhood residents to their governing boards and hired locals to serve on their staffs. They developed neighborhood improvement plans focusing on social services—job training, health clinics, legal aid offices, and preschool programs. The most important effect of introducing CAAs was that they did birth cohorts of activists, who, in subsequent years, would go on influence neighborhood and broader city politics (Bennett 2017).

At present, many cities all over the world have developed participatory procedures and structures that allow local resident comment on proposed land-use matters and, to some degree, mandate local resident advice on future private and public physical development within the boundaries of their neighborhoods.

Today the main driving force of neighborhood planning is localism. As I have mentioned earlier in this book, attachment to a particular place is a unique criterion defining a neighborhood and the foundation for decision-making at the neighborhood level (Park and Rogers 2015). As Brownill and Bradley (2017) note, neighborhood planning is constituted by localism.

The effects of localism can be seen across societies, with increasing devolution of power from central governments to local governments, including the establishment of new patterns of governance. The localism agenda seeks to empower communities by establishing neighborhood planning as a new, lowest level in the planning system. By giving legal weight to these plans, the local concerns and priorities become a formal part of the planning framework (Banfield 2020). In response to a persistent resident dissatisfaction with citywide planning outcomes, public authorities are transferring power to local communities as part of a wider decentralization agenda (Parker et al. 2019). Recent interest has grown dynamically as have become more aware of the benefits of place-based decision-making. Neighborhood planning is more responsive to local influences, as issues are small enough to effectively engage the residents and local stakeholders.

This shift in urban planning practices can be depicted with two very different approaches at the neighborhood level identified by Checkoway (1984). He called them "subarea planning" and "neighborhood planning." Subarea planning is initiated at the city level and involves the deconcentration of central planning activities to the neighborhood level, whereas neighborhood planning is community-based and involves the development of plans and programs by and for community residents themselves (Peterman 1999).

Another criterion for distinguishing subarea planning and neighborhood planning is citizen participation. Subarea planners are supportive of the aims of citizen participation, but their efforts do not lead to citizen control. They engage citizens only by using procedures of informing, consulting, and placating and claim that their jobs require objectivity and noninvolvement in political action. On the contrary, neighborhood planners reject the notion that planning can be neutral. Neighborhood planning objectives are not value-neutral, and it is the responsibility of planners not only to identify and articulate the specific values underlying planning prescriptions but also to affirm them. Whenever it becomes clear that the values of the planner and the community are at odds, the planner has an obligation to make these differences known and, if necessary, to withdraw from the project (Peterman 1999).

Neighborhood planning is also an educational experience. The most important thing a neighborhood planner can do is to use his or her knowledge and skills to help neighborhood residents understand the impacts of urban policies and obtain the resources a community needs to get involved in the process of decision-making. The ultimate goal is to teach the community and its leaders how planning and development are done so that when the planner leaves the community, the work will continue. This is the ultimate act of empowerment (Peterman 1999).

Practitioner's Perspective
What is the difference between urban planning and neighborhood planning?

Neighborhood planning is more granular than urban planning. When I think of urban planning, I think of whole cities or whole regions. When I think of neighborhood planning, I think more block by block. Neighborhood planning is just an ongoing engagement of a community, always occurring, always evolving process. Urban planning may happen in increments that are in years like 5 or maybe 10 years.

Patrick McMahon, Senior Vice President, Regional Development, Federal Realty Investment Trust, Somerville [interview conducted via Zoom 12/06/2021; reviewed and authorized by the interviewee on 11/19/2021; origin: Fulbright Senior Award scholarship 2021]

The difference between urban planning and neighborhood planning is in scale but not in substance. The difference is the amount of information and feedback you have to coordinate in order to make sure that all members of the community have the opportunity to share their particular expertise.

Rebecca Lyn Cooper, Senior Planner, Planning & Zoning Division, Mayor's Office of Strategic Planning & Community Development, Somerville [interview conducted via MS Teams 11/19/2021; reviewed and authorized by the interviewee on 11th of December 2021; origin: Fulbright Senior Award scholarship 2021]

To recap all the aspects discussed above, it may be stated that neighborhood plans are policy documents that identify the tasks that need to be carried out to improve a given area of a city and provide a particular recommendations and guidance for development. They are based on an analysis of a large amount of data collected about that area, and, if prepared correctly, they represent a consensus among the stakeholders about the future vision of the neighborhood (Jones 1990). A core requirement is that the neighborhood plan must be in a broad conformity with the relevant strategic aims of the municipality. As a result, it may be unable to refuse some strategic development allocations that the residents do not want (Banfield 2020).

In its long history, neighborhood planning has significantly reconfigured the power relations between citizens, the state, and the development industry (Brownill and Bradley 2017). As Peterman (1999) postulated, neighborhood planning is more than the physical design of local space: it involves community building, identifying not only the local needs but also the ways in which people link with communities beyond some limited administrative boundaries. This final observation points to the concept of neighborhood governance characterized in the next section.

2.10 Neighborhood Governance

In general terms, governance is about working together, about the cooperation between different stakeholders (Somerville et al. 2009). Kooiman (2005) has distinguished three modes of governance: hierarchical governance, self-governance, and co-governance. The first one is a "top-down" decision-making model, performed by a "governator" who represents a particular community. Self-governance is a "bottom-up" approach where a collectivity is able to shape its decisions itself. Co-governance is then where a group cooperates with other groups in a process of mutual shaping and decision-making processes.

Following this division, neighborhood governance can be hierarchical, conducted by some institutions based outside the neighborhood, where the leader is a governmental body, implementing particular policies within the neighborhood. A neighborhood can be also self-governed, when the residents themselves shape and represent their community. Here, governance institutions are some kind of associations or forums based in the neighborhood. Eventually, it can be also a co-governance model, where the shaping and representation is done jointly by governmental and nongovernmental bodies in the neighborhood (Somerville et al. 2009).

Neighborhood governance is also a capacity to establish the relations with others. It can be achieved in a spontaneous, unplanned way among individual residents, or it can be planned through collective bodies set up specifically for this purpose (Somerville et al. 2009). As the number and diversity of actors in urban policy is growing (Belof 2013), also the variety of forms of local governance is increasing. Measures to grant the local stakeholders a genuine empowerment and give them greater responsibilities need to be adopted, extending the range of governance on those who are excluded or oppressed in some way.

And here is the difference between neighborhood planning and neighborhood governance. While the planning system is conceived in vertical terms, the local governance is often conceived horizontally as a patchwork of spaces over which different civic bodies have responsibility. And even if crude distinctions between vertical and horizontal conceptions of decision-making are rather simplistic, as these structures are not static but are liable to change, a new conceptualization is needed to understand local planning and governance. As Banfield (2020) notes, a networked approach to space moves us one step closer, as it does not assume orderly, static, or simplistic spatial structures but instead emphasizes multiplicity, contingency, and messiness. Such network-understanding of neighborhood governance brings multiple interested parties with different perspectives together to develop planning policies, to reduce conflicts and speed up development by reducing the number of appeals against planning documents.

Neighborhood governance also involves the organization, monitoring, and delivery of goods and services, the maintenance and enforcement of reasonable standards, and the provision of acceptable environmental conditions within agreed lines of control and accountability. It cuts across traditional hierarchies and forces local policy practitioners into a much more exposed position, closer to the "front line."

Their responsibilities include (1) dedicated budget and spending power with the local delivery team, (2) direct delivery of core services, (3) coordination of wider service inputs, (4) direct responsibility for resident consultation and involvement, (5) housing management and many related services, and (6) responsiveness to council (municipality) executives (Power 2004). All of those functions should be carried out effectively by neighborhood governance in order to fulfil the day-to-day needs of the local residents.

2.11 Integrated Planning

As it has already been stated in the introductory section of this book, laying the foundations for a novel concept should be preceded by a thorough overview of the existing ideas in a given field. Therefore, after defining the core characteristics of a neighborhood (including its functional program, the relation between the community and a particular area, as well as the spatial and behavioral boundaries), presenting various descriptive approaches (neighborhood cohesion and livability), and outlining the definitions of neighborhood planning and governance, below several other approaches to urban policy which seem to be noteworthy in the context of conversion and hybridization of urban services have been discussed.

The first is integrated planning. It aims at coordination of various policies in order to produce comprehensive development strategies and may take various forms: institutional, indicating particular standards, procedures, and people in charge, or simply practical, encompassing good administrative practices. The concept of integrated planning implies cooperation between various actors and disciplines, offering a holistic decision-making framework addressing all the major challenges in territorial development (Markowski 2011; Gzell 2013). It should take into account both local and regional policy objectives, it should be responsive, and it should adjust long-term and short-term plans to particular ever-changing territorial contexts (Ministerial Meeting 2011; European Commission 2013).

Integrated approach is essential for successful policy outcomes which has been proved in many European area-based policies (including neighborhood development strategies). The advantage of integrated efforts is clear: overlapping problems can be dealt with from different angles. However, the success can be guaranteed only when all parties (e.g., local government, housing associations, shop owners, inhabitants, private developers) come to an agreement about the content of the policy (Somerville et al. 2009).

The great popularity of the concept of integrated planning stems from its multilayer approach. However, its far-reaching perspective makes it to some extent an abstract and vague term. Literally any collaborative or interdisciplinary activity undertaken by public institutions and related to spatial development may be labeled "integrated planning." As a result, it seems to be just a new name for the processes that would have happened anyway and does not provide answers to specific shortfalls of traditional urban planning. One of the phenomena that elude integrated planning framework is online conversion of services which is hard to grasp taking purely the statistical, data-driven approach to policymaking.

2.12 Smart City

The smart city concept is a part of a wider trend of smart growth, popularized all over the world in the early 2000s. It offers a prospect of a society that is not only environmentally sustainable but also sufficiently competitive and cohesive to meet the quality-of-life agenda.

Smart city approach is characterized by information-rich networks of innovation and highly communicative, creative partnerships (Deakin and Waer 2012). It is thus relevant to deal with conversion of services phenomena (more than integrated planning described above). Smart city governance aims at coordinating urban transport, environmental issues, and economic development; it synthetizes the "hard infrastructure" (urban networks and facilities) and "soft infrastructure" (knowledge, communication, and social capital), the latter being critical for a city's competitiveness (Leboreiro Amaro 2014). As a result, it assumes a harmonious urban future, based on the growing role of human capital, social learning, and high-tech communities (Hollands, 2008). Having strong institutional support, many cities all over the world have adopted particular "smart policies" to promote themselves as forward-looking and prosperous (Deakin and Waer 2012).

But SC should not be perceived only as an attractive label for contemporary urban areas. According to the European Commission (2013), smart cities should be seen as systems of people interacting with each other, where flows of information, energy, materials, services, and finance promote sustainable development, resilience, and a high quality of life. The "smartness" of these flows and interactions comes from communication infrastructures used in strategic urban planning and management. In other words, ICTs are increasingly pervasive, making the economy and governance progressively driven by knowledge, innovation, and entrepreneurship. Another important feature of smart development is transparency: smart cities are responsive to the social and economic needs of society (JPI Urban Europe 2015) and aim at providing high level of participatory, democratic decision-making processes, also based on the latest ICT solutions.

Paradoxically, smart governance provides only a partial response to the online conversion of services. While it promotes the implementation of online solutions in the public sector (i.e., services of general interest, including administration, education, or healthcare), it overlooks the possible negative impacts of digitalization on urban areas (such as entrepreneurship that evokes public spaces). It seems clear that contemporary urban governance should consider and address the aspatiality of online services, but the smart city concept seems to overlook this issue (Damurski 2021).

And, of course, the development of the concept has been followed by some criticism. The so-called human smart cities manifesto (Balducci 2013) and other similar works (e.g., (Deakin and Waer 2012)) advocated the alternative, humanistic perspective on the use of ICT. According to this approach, smart cities should not be hardwired but should be socially inclusive, introduce good governance, and create services that focus on improving the quality of life. A progressive city cannot be labeled "smart" simply by adopting a sophisticated digital infrastructure—it also requires the input of different groups of people. Despite these objections, SC is a concept that has gained real momentum and remains one of the core strands in the urban governance discourse.

2.13 E-planning (Basics)

E-planning was a concept firstly developed by researchers publishing their works in *International Journal of E-Planning Research*. Linking the online and offline realities in the e-planning paradigm seems to be highly relevant for addressing the online conversion of services.

Definitions of e-planning use different criteria and therefore range from very broad to relatively narrow. In the broadest sense, e-planning is a sociocultural and political practice in which people participate both in-person (offline) and remotely (online) according to the subsequent stages of the planning process (Horelli and Wallin 2010). This definition captures all forms of social interaction and decision-making, suggesting that e-planning is a holistic term for everything that has traditionally been called "planning." However, it does not explain how e-planning differs from what we have known before (Damurski 2021).

A narrower understanding of EP focuses on selected aspects of urban planning processes, such as citizen participation (Andrzejewska et al. 2007), online interfaces (Antunes et al. 2010), or modes of spatial decision-making (Damurski 2021). Such sectoral approaches give a good insight into how contemporary spatial governance takes advantage of numerous technologies to perform its tasks, including powerful web-based systems such as spatial information systems. Building online interactive questionnaires for citizens to complete administrative procedures and incorporating social media into public communication strategies for increased transparency and instant flow of information between different stakeholders—systematized and integrated—are often referred to as Planning Support Systems (PSS), which combine the complex spatial analysis and display capabilities of GIS with personalized, on-demand, customer-centric support.

Following the theoretical debate of the previous decades (Budthimedhee et al. 2002; Szuba 2006; Klosterman 2012), e-planning can be generally defined as a public decision-making process related to particular territories, enabling the communication between various stakeholders and the planning institutions with advanced ICT applications, respecting the principle of equality of all actors (Damurski 2021). Hence, EP to some extent illustrates the vital interdependence between communication technologies and spatial planning, including the aspects of territorial management, policymaking, governance, citizenship, and participation (Curwell et al. 2005).

To sum up, it may be noticed that the new governance practices greatly depend on ICT infrastructures. When online connectivity of a society is low, then the potential of e-planning cannot be fully exploited; similarly, when planning culture remains authoritarian, then even the best e-planning facilities will be useless. These wide sociopolitical and technological contexts highly impact the way e-planning is understood and implemented and affect the manner in which traditional urban policy is being transformed (Wallin, et al., 2012). Therefore, when studying the emerging e-planning practices, we need to take into account the character of planning systems, the condition of societies, and the dominating policy practices.

2.14 Urban resilience Agenda

Accelerating diversification of urban functions and massive migrations to urban areas in many parts of the world increase the risk of various kinds of disasters endangering cities: natural, humanitarian, or economic. Building and enhancing resilience to various disruptions has recently become a critical issue in urban governance. Local authorities regularly assess the vulnerability of urban infrastructure subsystems and develop risk management strategies and disaster scenarios in order to minimize the impact of potential disruptions (Kapucu et al. 2023).

Resilience means a relative stability in providing a certain standard of urban life and a continuous process of growth, transformation, adaptation, and renewal (Lang 2010). A resilient urban system can be defined as one that can withstand external shocks and absorb extreme stress while keeping its primary functions, even if they are performed in a changed form (Innes and Booher 2010).

The concept of resilience initially referred only to environmental systems, but due to the growing complexity and nonlinearity of urban development, it is widely acknowledged that urban resilience concept should be located in system thinking instead of focusing on a single dimension. The definition of urban resilience proposed by Meerow et al. (2016) seems to be the most accurate one. It states that urban resilience is the ability of an urban system (and all its constituent networks across various temporal and spatial scales) to maintain or rapidly return to desired functions in the face of a disturbance. It also involves adaptation and quick transformation of subsystems that limit current or future adaptive capacity (Meerow et al. 2016).

As Zhang and Wang (2023) note, resilience indicators correspond to the main urban subsystems: infrastructures, socioeconomic system, governance, and energy and material flows. Urban policy and management practices indicate the indispensable role of information and communications technology in improving urban resilience (Zhang and Wang 2023). ICT is an element of critical infrastructure supporting the everyday functioning of cities and their ability to react to the emerging tensions. Moreover, in the long run, the deployment of digital solutions helps in ensuring that cities are more resilient also regarding the climate change: by reducing the need to travel, it contributes to declining emissions, linked to a reduction in automobile use (Moreno et al. 2021).

But urban resilience is not only about the technical resources and infrastructures that municipalities have at their disposal but also about the way the decisions are taken—what elements they involve and how effective they are. And this is the domain of urban governance. An adaptive capacity is required in managing cities, including better understanding of changes occurring in complex, interdependent social and spatial systems as well as developing the practices that will allow to react quickly to different situations (Innes and Booher 2010; Lang 2012).

The recent COVID-19 pandemic (2020–2022) and the resulting global lockdown heavily impacted food, accommodation, livelihoods, public transport, economy, and other public amenities in cities (Chepelianskaia 2020). Many supply chains have been broken, and many traditional services providers all over the world faced severe challenges, including organizational and financial threats. Simultaneously the pandemic has shown the great potential of the Internet in building resilience of urban areas: online services compensated the shortages of spatial availability of various facilities and enabled retaining the supply of basic products for individual residents and for the whole communities (Damurski 2021).

The unprecedented global health crisis has put the issues of urban resilience high in the agenda of public policy. It also provided new arguments for developing a multichannel approach in urban governance that would mix "online" and "offline" solutions and thus contribute to the overall stability of urban systems. Innovative hybrid policies need to be developed both in relation to the content of urban governance (land use and services provision) and its procedures (citizen participation and public communication).

2.15 The 15-min City: Proximity, Accessibility, and Walkability

Households meet their needs in a range of different types of urban centers, following the economics of demand and supply (the dynamics of concentration mechanisms and agglomeration effects will be further discussed in the following chapter). The complexity of spatial interactions between services supply and residents' demand is a vital condition in developing "compact city" planning approaches. It cannot be assumed that all needs can be met locally; however, the balance of activities across different types of areas can be achieved through urban land-use management (Fairgray 2012). This is why in compact cities, where high density of housing is accompanied with various facilities, residents appear to be significantly more satisfied with their neighborhoods compared to residents of sprawled suburbs (Mouratidis 2018).

Box 2.3: COVID-19 pandemic and the online retail in the USA

Over the past 20 years, e-commerce has steadily risen in size and reach. In 2019, e-commerce sales accounted for just 11% of total retail sales (excl. Food services), while the COVID-19 crisis accelerated the ongoing shift to online retail. Faced with stay-at-home orders and store closures, millions of Americans resorted to shopping online, resulting in a three-point jump in the online portion of total retail sales. According to the US Census Bureau, e-commerce sales amounted to $792 billion in 2020, which is equivalent to 14% of total retail sales. While 14% still doesn't seem like a lot considering the perceived role of online retail, it needs to be noted that total retail sales include categories such as motor vehicle and parts dealers, gas stations, and of course grocery stores, where e-commerce still plays a very minor role. Other categories, most importantly clothing and footwear, are seeing significantly higher shares of online sales already (Richter 2021).

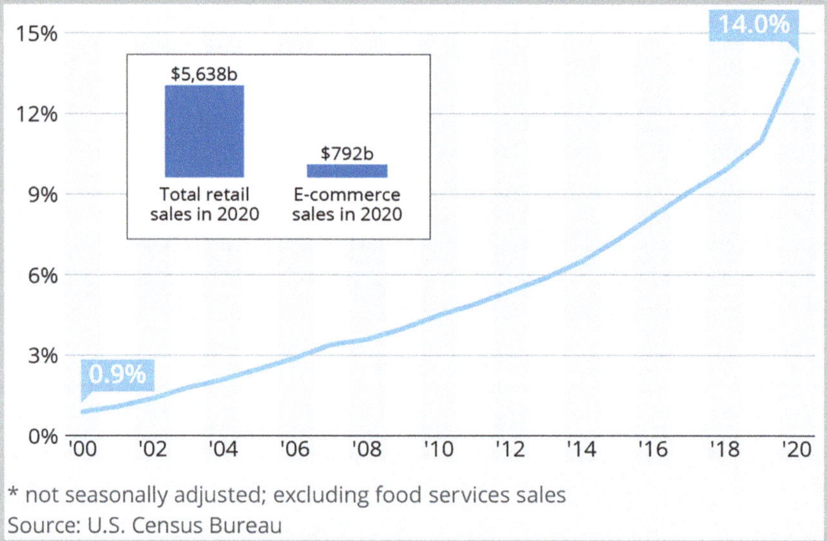

Chart showing the dynamics in e-commerce in the USA in the years 2000–2020. (Source: Article by Felix Richter published on https://www.statista.com/chart/14011/e-commerce-share-of-total-retail-sales/ under the Creative Commons license Attribution 4.0 Attribution-NoDerivatives 4.0 International (CC BY-ND 4.0) https://creativecommons.org/licenses/by-nd/4.0/. Retrieved on 15/09/2023. No changes made. Origin: Fulbright Senior Award scholarship 2021)

To create a vibrant urban area, full of pedestrians, increased face-to-face contacts, and feeling of security, the idea of "15-min city" is worthy of embracing (cf (Park and Rogers 2015)). The concept emanates from the "chrono-urbanism" and promotes increased proximity and social interaction. It is a follow-up of the traditional ideas of human-scale urban design (Alexander et al. 1977a, b; Gehl 2013; Whyte 1990). The quality of urban life is inversely proportional to the amount of time spent by the residents on transportation; therefore, urban structures should enable locals to access all of the basic facilities at distances that would not take more than 15 min by foot or by bicycle (Moreno et al. 2021).

The classical version of "15-min city" postulated that residents—in order to sustain a decent urban life—should be able to effectively perform six essential urban social functions: (a) living, (b) working, (c) commerce, (d) healthcare, (e) education, and (f) entertainment. To attain those functions, the urban tissue needs to ensure proximity, diversity, density, and ubiquity.

The revision of the concept presented by Moreno et al. (2021) added some new components relevant for the current era, including the ubiquitous deployment of ICT. With these technologies, cities can enhance and improve service delivery as well as promote sustainable practices. The "modified 15-minute city" framework depicts four dimensions: (a) density, (b) proximity, (c) diversity, and (d) digitalization. These were identified after observing the challenges that different cities across the globe endured during the COVID-19 pandemic and the subsequent restrictions aiming at mitigating the spread of the virus, as well as the numerous challenges in accessing basic facilities by the residents (Allam 2020).

Regarding proximity dimensions, digitalization proves to be effective where online services, cashless transactions, and virtual communications are implemented and promoted. Since the COVID-19 pandemic, it has been possible for people to work from home and interact virtually, and this also has been instrumental in reducing the need to commute from home to offices and other work-related places. Providing some services within the comfort of homes also contributes to the goals of "15-minute city" and is crucial in optimizing the consumption of resources. For those reasons, digitalization is expected to stay in the post-pandemic urban planning jargon (Moreno et al. 2021).

The concept of proximity is directly connected with other terms describing the local services system: availability, accessibility, and affordability. In short, availability deals with the range of services offered in a particular location. It is modulated by the local market: the relationship between supply and demand determines the types of services offered in a neighborhood. Accessibility describes the way people penetrate the area on their way from home to various services, using various modes of mobility. Eventually the term affordability is connected with the level of affluence of the local community. It is used to answer the question whether the services offered in a given neighborhood are not too expensive to the clients. All those perspectives will be described in detail further in the book.

Another trend in urban planning, conceptually linked to the idea of the 15-minute city, points to the need to design pedestrian-friendly environments and to increase the "walkability" of urban areas (see, e.g., (Zhang and Mu 2019)). In addition to improving public health, walking reduces traffic congestion, energy consumption, air pollution, and associated costs. It also offers more livable communities with shorter trips and economic benefits to the local business. Walkability can also reduce the inequality of access for anyhow disadvantaged people. Walking is not only a mode of transportation: it can be a social choice, a recreational exercise, or even an aimless activity (Litman, 2018).

However, in the perspective of the progressing virtualization of human daily behaviors, walkability can be viewed ambivalently. On one hand, it seems to be necessary to provide an attractive alternative to home-centeredness, offering an opportunity for routine, everyday walking to basic services as well as additional in-person social interaction in a neighborhood. On the other hand, however, even high walkability indicators may not be sufficient to encourage residents to move around the area when most products can be ordered online and delivered door-to-door. Therefore, the pedestrian-friendliness mentioned above should be followed by various spatial and functional solutions offering a viable alternative to virtual reality in order to motivate the residents to change their habits and start walking through their neighborhood again.

2.16 Intermediate Conclusions

Neighborhood is one of the essential concepts in urbanism studies, a building brick of contemporary thinking about urban areas and urban governance. It was popularized in the early twentieth century and was expected to provide a suitable living environment within a given area and social cohesion within a given population. Following a particular division of the city into smaller subunits, the neighborhood gained political power and became an important player in the local decision-making processes.

Today, a neighborhood is perceived as a group of people who are characterized by a certain degree of social cohesion and who share particular services in close proximity to their homes, within a clearly delineated subarea of a given city. For the purposes of this book, which aims to capture the neighborhood phenomenon in holistic way, the term is defined twofold: (1) as an administrative unit where particular political power is exercised through the formulation of territorial policies and (2) as a local functional area where essential public and private services are provided, a place of everyday social practices of the local community. Both of those meanings need to be incorporated into the proposed idea of multichannel governance.

Hence, the vital interplay between geographical and social features distinguishes neighborhood from other types of urban communities. Paradoxically, this is also the reason why neighborhood becomes a transient concept. Both territorial and social functionalities are currently under pressure of new technologies. As a result of growing, immersive pressure of the ICT, we are entering the times of transient space and transient society. Consequently, a more relative, abstract approach to neighborhood is needed.

The overview of urban governance paradigms and policy approaches presented in this section enables drawing a comprehensive picture of the challenges faced by urban neighborhoods. These are as follows:

- Monofunctionality resulting from spatial structure focused merely on residential land uses.
- Dispersion and decentralization of nonresidential urban functions resulting in limited access to essential, everyday services.
- Car dependency in commuting between homes, workplaces, and services.
- Social isolation of individuals within the local community.
- Limited connections with other urban areas resulting in the isolation of a neighborhood.
- Weak ties to the place of residence resulting in lack of local identity.
- Selectivity of citizen participation activities, involving only the well-informed residents and resulting in political exclusion of other groups.

Those challenges are followed by various spatial, functional, and organizational solutions aiming to mitigate the potential threats and risks. These are as follows:

- Stimulating livability and self-sufficiency, including access to essential daily services, public spaces, as well as green, open areas.
- Fostering social cohesion and behavioral and functional ties between groups and individuals.
- Building and strengthening local identity and place attachment.
- Reducing the need to travel by enhancing mixed land-use patterns and merging residential functions and workplaces followed by high walkability.
- Supporting connectivity with other urban areas using various means of mobility, including public transport.
- Providing "third places" for direct, in-person interaction and socializing.
- Promoting centrality values by building local service hubs where various neighborhood activities would concentrate.
- Developing effective citizen participation in local governance, with special attention to groups with limited access to information.

Technological aspects of neighborhood governance have been covered only in some of the policy concepts presented in this chapter. First, selected urban planning manifestos indicate to the impact of the ICT on performance of urban areas, particularly regarding the availability of information, stakeholder communication, social interaction, work-home relationship, and land-use structures. Second, the smart city concept is essentially connected with digital technologies, implemented both in managing urban infrastructures and in building human and social capital of urban communities. Third, e-planning practices are developed by adopting the ICT in all forms of contemporary decision-making processes, including particularly citizen e-participation. Fourth, urban resilience involves building communication infrastructures as a critical component of city's ability to react to unexpected disturbances. Fifth, the revised 15-minute city concept introduces digitalization as a way of overcoming shortages in services provision and workplaces accessibility.

All of those attempts to respond to the challenges of the ICT development with urban policy seem to be quite random, ad hoc, disperse, and shortsighted. They do acknowledge the importance of digital tools but limit their influence only to some selected sectors of urban governance. Moreover, they do not address the specificity of a neighborhood and thus lose their potential at the local scale. Hence, the need to develop a holistic vision of multichannel governance becomes urgent and necessary.

The subsequent chapters of this book will follow the remaining stages of cognitive sequence presented in the introduction. Current development of the virtual urban environment will be presented in two aspects: multichannel objects of neighborhood governance and multichannel processes of decision-making. As a result, in the concluding section of the book, a concept of multichannel neighborhood governance will be introduced, including the necessary revision of to-date paradigms of urban policy.

Chapter 3
Multichannel Objects of Governance

Abstract In this chapter, various civilizational impacts of the ICT development are presented, followed by an in-depth analysis of the urban service sector as the main area where virtualization effects can be observed. The study adopts social, geographical, and political perspectives to demonstrate the profound and multifaceted consequences of information and communications technologies on urban development. Its main argument is that the shift from traditional services to online services in the recent years has a considerable impact on neighborhoods which requires a targeted intervention from urban governance.

Keywords Metaverse · Virtual territoriality · Internet generation · Concentration · Urban services · Accessibility · Conversion · Hybridization · Neighborhood services · Multichannel services

3.1 Time-Space Compression and the Death of Distance

The overview of various perspectives to urban local policy presented in the previous chapter has to be complemented with a wider reflection of the ICT impact on urban structures and neighborhood communities. It is a necessary element of laying foundations for multichannel neighborhood governance.

One of the first concepts addressing the changing links between time and space in urban areas was the "time-space compression" introduced in the late 1980s by D. Harvey (1989). It focused on the reduced need for traveling observed at the end of the twentieth century. While most of the human activities (work, commerce, education, healthcare, administration, entertainment, etc.) can be performed remotely, the only constraint in the process of endless "aspatialization" is the access to high-speed transmission networks (Harvey 1989). A slightly more explicit vision of the online-offline relationships presented by Cairncross (1997) proclaimed even the "death of distance." As a result of the growing ease of communication, the physical

Ł. Damurski, *The Foundations of Multi-channel Neighborhood Governance*, Local and Urban Governance, https://doi.org/10.1007/978-3-031-75124-0_3

distance does not affect the human's ability to work or interact with others anymore, highlighting the need to revise our perception and understanding of space.

From a geographical point of view, the increasing digitalization of the economy makes physical localization less important than in the past. Industrial production is being replaced by new forms of nonmaterial production based on knowledge, creativity, innovation, business, and cultural and leisure functions in urban areas (cf (Batty 2014)). The advantages of physical location are being replaced by greater attention to citizens' human capital (knowledge and skills) and access to high quality communications infrastructure (Sassen 2010; Llewelyn-Davies et al. 2004).

Hence, today's cities exist simultaneously offline and online, becoming a hybrid formation of two parallel constructs: physical (embedded in the geographical space) and virtual one (comprising of online worlds and interactions). A simple observation of the "visibility" of urban areas in the Internet proves that their online positioning fairly represents their real counterparts (Janc 2014). As a result, the ICT development may—to some extent—threaten the very existence of the city as a center of human life and activity (Hall 2002). While the Internet takes over some well-established urban functions such as commerce, finance, or public administration, we might witness an approaching collapse of the city, or at least of urbanity as we have known it for centuries (cf. (Castells 1998)). Finding the right balance between the real and the virtual seems to be one of the major challenges in urban governance since the deindustrialization era (Palej 2005).

Thus, today's cities exist simultaneously offline and online, becoming a hybrid formation of two parallel constructs: physical (embedded in geographical space) and virtual (consisting of online worlds and interactions). A simple observation of the "visibility" of urban areas on the Internet proves that their online positioning fairly represents the ranks of their real counterparts (Janc 2014). As a result, the development of ICT may—to some extent—threaten the very existence of the city as a center of human life and activity (Hall 2002). While the Internet is taking over some well-established urban functions such as commerce, finance, or public administration, we may be witnessing the imminent collapse of the city, or at least of urbanity as we have known it for centuries (cf. (Castells 1998)). Finding the right balance between the real and the virtual seems to be one of the major challenges of urban governance since the deindustrialization era (Palej 2005).

3.2 Virtual Reality and the Metaverse

The concept of the metaverse has recently gained a growing attention among scholars all over the world. It is a term taken from the science-fiction literature, which indicates a virtual universe parallel to the real world within which individuals can move, share, and interact through personalized online applications. The etymology of the word derives from Macedonian, formed by the prefix meta- (inside) and the noun verse (abbreviation of universe) (De Felice et al. 2023).

The metaverse can be defined as a "universal, persistent and immersive virtual world, accessible through virtual and augmented reality devices, in which users can be involved in diverse activities" (Vidal-Tomás 2023, p. 1). In other words, it is a shared online space characterized by a seamless technological mediation, which allows user-generated digital characters to interact with each other in a real-time 3D environment (Yoo et al. 2023).

The development of the metaverse strongly depends on technologies that drive the transition from the current Internet. Several tools are necessary for popularizing the metaverse: mobile networks, extended reality, user interactivity, artificial intelligence, blockchain, Internet of things, robotics, as well as cloud computing (Vidal-Tomás 2023). Each of them provides a specific contribution to the real-virtual connectivity.

There are many giant companies investing in metaverse. Meta (previously Facebook) is the main one, but there are also other similarly large tech actors such as Microsoft and Nvidia. Hundreds of other multinational firms and institutions are their partners (e.g., The Sandbox and Decentraland), constituting a fast developing sector of contemporary global economy (Vidal-Tomás 2023).

Referred to as the "new Internet," the metaverse aims to become a virtual world where people can have fun, socialize, and work. It involves strong interaction between digital and material environments and arises particular questions for the future: will virtuality completely replace reality? How can humans switch freely between virtuality and reality not losing a psychological balance? These challenges need to be considered in the neighborhood context if we aim to manage it in a successful way (Shi et al. 2023).

3.3 Changing Behaviors

User experiences in the real world are restricted by spatial-temporal coordinates. Thoughts, feelings, and all kinds of human interactions in daily life occur at a given time and place, depending on actual context, and are abided by some obvious rules (e.g., speed of traveling has upper limitations, past time cannot be called back, etc.) (Shi et al. 2023). Metaverse is capable of going beyond the simply three-dimensional world, especially in the aspects of space and time, combining the physical and digital environments (De Felice et al. 2023), and one of its key features is a sense of direct interaction with a virtual world: impressions gained from metaverse become similar to those from the real world (Oleksy et al. 2023).

Supported by virtual reality, the metaverse offers a mix of spatial and social presence. Users perceive others as being physically close to them, sharing the same space. They can interact with the simulated environment, as well as with the avatars of other users (their digital representation), and all this is designed in a very intuitive way: users can approach and talk to each other like in the physical world. Hence, the feeling of being physically placed in a virtual space (spatial presence) in the company of others (social presence) enables relatively intimate and close relationships with people and advocates for the metaverse as an attractive alternative to the real world (Barreda-Ángeles and Hartmann 2022).

Practitioner's Perspective
Virtual reality experience

When I use social VR platforms, I feel like I am actually there in the virtual environment (...). It is as though my true location shifts into the virtual environment.

Example of items used in the self-location subscale of the Spatial Presence Experience Scale [source: (Hartmann et al. 2016)]

The popularizing virtual reality is changing various behavioral patterns of people. It is particularly visible in the group of teenagers who spend a lot of time texting on their phones, using social media, surfing online, and gaming—but not talking to each other. Their phone is usually the last thing they see before they go to sleep and the first thing they see when they wake up. This phenomenon has a profound impact on their lives, including everyday social and spatial behaviors.

An extensive research conducted on American teenagers by Twenge (2017) shows that their characteristics are distinctly different from their predecessors. The oldest members of the so-called iGen (people born between 1995 and 2012) were early adolescents when the iPhone was first released in 2007 and became high school students when the iPad was introduced in 2010. The important trends shaping iGen'ers are extending childhood into adolescence, large proportion of time spent on mobile phones, decline of social contacts performed in-person, rise in mental diseases, decline in religion, focus on individual safety, reduced civic engagement, new attitudes toward income and work, loose approach to relationships, tolerance, and political independence (Twenge 2017).

Some authors suggest that metaverse fosters the emergence of a new society (cf (De Felice et al. 2023)). Indeed, the immersive ICT may play an important role in social connectivity and users' well-being (especially in the periods of limited face-to-face contact, e.g., during the COVID-19 pandemic). Metaverse enables activities like chatting with friends, meeting new people, playing games or attending virtual concerts, exhibitions, etc. At the same time, social virtual reality platforms address and satisfy the needs that may have never been met within the offline world. They allow users to create and edit creating new contents (new things, virtual worlds, avatars, etc.) and provide a common space that merges the physical and digital worlds (Oleksy et al. 2023; Barreda-Ángeles and Hartmann 2022).

Therefore, "where iGen goes, the country goes" (Twenge 2017, p. 15). The new generation brings their values and viewpoints to the market: their lifestyle and related preferences are already shaping the commercial trends in many parts of the world. Their social behaviors affect the demographic characteristics of a growing number of countries, raising the question whether there will be enough workers to support previous generations in their retirement. iGen is definitely at the forefront of the profound civilizational changes taking place today.

Box 3.1: Changing Teenagers' Habits

iGen'ers (people born between 1995 and 2012) spend less time interacting with their peers face-to-face than any previous generation. It is not just parties or craziness but merely getting together with friends, spending time hanging out. They are seeing their friends in-person 1 hour less a day than Millennials did when they were their age. An hour a day less spent with friends is an hour a day less spent building social skills, negotiating relationships, and navigating emotions. It has been replaced with screen time (Twenge 2017).

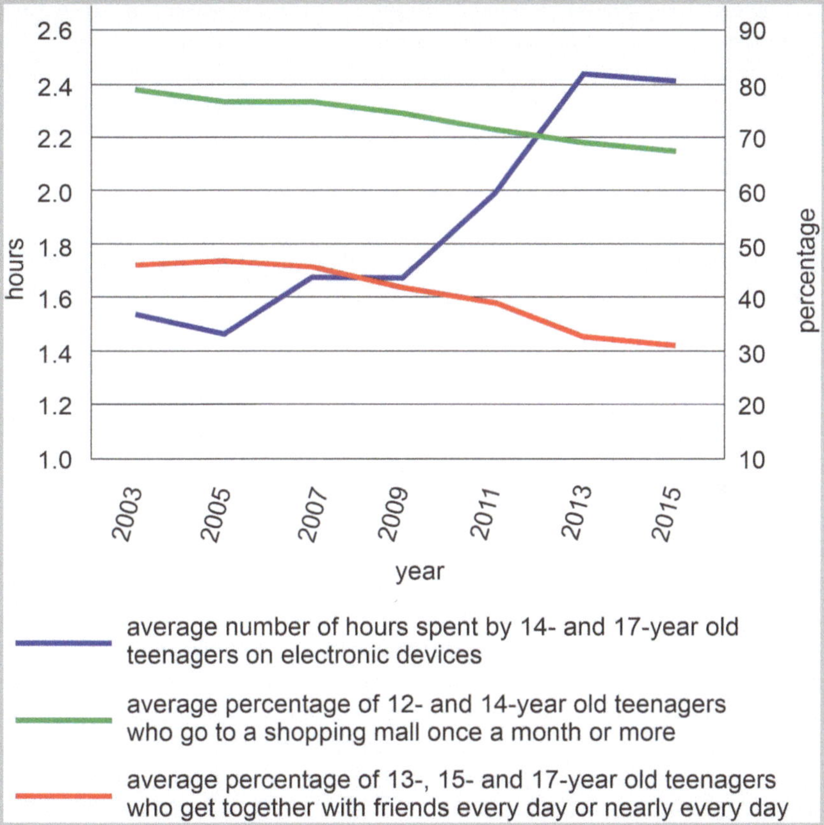

The dynamics of selected characteristics of iGen. (Source: Author's own research based on the data presented by Twenge (2017). Origin: Author's own research)

To recap this section, we need to note that the Internet generation is also a growing group among stakeholders and decision-makers. The iGen'ers' worldviews, political preferences, and lifestyle will soon shape urban discourse, having profound impact on policy priorities, directions, and practices. This is one of the reasons for revising the neighborhood governance paradigms and adapting them to the changing values of contemporary society.

3.4 Virtual Territoriality

Fading in-person social interactions include everything from one-to-one activities such as getting together with friends through small group gatherings to parties in larger groups. They embrace activities with no real aim (such as cruising in a car) and those that have a particular goal (such as going to the cinema) (Twenge 2017). All of those traditional socializing opportunities are being replaced with online ones, reducing the role of space in everyday human behaviors. As Castells (1998) puts it, people more and more often work and consume services at home, thus contributing to the *homing* trend. And this direction will keep on to speed up as long as new technologies are being introduced.

The feature that distinguishes the metaverse from its predecessor, the Internet, is its capacity to be explored in a manner analogous to how people interact with real places. This effect is achieved through both virtual and augmented reality technologies that can either transfer users to a virtual location or integrate virtual elements into the real world (Oleksy et al. 2023). As a result, human territorial behaviors are significantly altered: on one hand, we do not need to move from home to satisfy various needs, but on the other hand, our homes have to be specifically equipped to meet the technological requirements of the metaverse.

In order to address this complex issue, we first need to note that physical places have formed natural conditions of human existence for ages. The classical definitions of a place assumes that it is a bounded entity with a unique identity and historical continuity. Other definitions stress that a place should instead be defined as a location with interactive potential, a meeting space accessible to all members of a community rather than a separated enclave. However, most researchers agree that a place, as opposed to a space, possesses a specific meaning for its users (inhabitants or visitors). Thus, a place is not a random location which we just pass by but an object to which people are connected through various individual and social life experiences (Oleksy et al. 2023).

A term which seems to be relevant and useful to describe the relationship between humans and spaces is territoriality. Human behavior has a natural tendency to control the resources within a certain area. This strategy, implemented by both individuals and communities, involves creation of borders and zones for modulating social relations (Sack 1986). Territorial references play an instrumental, symbolic, and emotional roles in human spatial behaviors. The place of residence, its surroundings, and public spaces are treated by humans as their "own," despite their varied formal ownership (Jałowiecki and Szczepański 2002), which usually results in some kind of attachment and loyalty to particular geographical area. All this is observable in various ordinary and occasional behaviors intimately related to a certain place, starting from daily shopping and ending with periodical local cultural events.

Practitioner's Perspective
We do not socialize in real anymore

People stay in more often. My generation lost interest in socializing in real—they don't have physical get-togethers; they just text together and just stay at home.

Kevin, third grader in high school in San Diego, CA [source: Twenge, J. M., 2017. iGen. Why Today's Super-Connected Kids Are Growing Up Less Rebellious, More Tolerant, Less Happy - and Completely Unprepared for Adulthood. New York, London, Toronto, Sydney, New Delhi: Atria Paperback, p. 69; origin: author's own research].

The last generation always wants us to be in person, and a lot of us are not like that. We're more of a technology-based generation.

Darnell, 20-year-old college student in Georgia [source: Twenge, J. M., 2017. iGen. Why Today's Super-Connected Kids Are Growing Up Less Rebellious, More Tolerant, Less Happy - and Completely Unprepared for Adulthood. New York, London, Toronto, Sydney, New Delhi: Atria Paperback, p. 76; origin: author's own research]

Another relevant term is place attachment, often defined as an emotional, cognitive, and behavioral bond developed by individuals and communities with places (Lewicka 2011). This bond can stem from everyday life experience, the period of residence in particular area, and the cultural (or historical) significance of a place. Attachment brings particular benefits for human lives as it determines well-being and enables stress reduction and the feeling of belongingness. Moreover, a place can play an important role in maintaining self-esteem and self-efficacy: for some individuals, living in a particular area is considered a source of pride (Oleksy et al. 2023).

The emergence of virtual worlds has raised unprecedented challenges regarding the role of physicality in defining meaningful locations. Traditionally, places have been conceptualized by their tangibility and a concrete physical basis. But the developing digital worlds have a comparable potential to provide comfortable spaces for people: virtual environments can be easily accessed via desktop or mobile devices, presenting an attractive, alternative universe that looks very real (Oleksy et al. 2023). Moreover, as Skolik (2015) argues, certain aspects of territoriality can also be found in virtual space. Some online sites are treated by particular Internet communities as their own, resulting in behaviors which are typical for offline territoriality, such as appropriating certain virtual places, establishing borders, and defending them (Skolik 2015).

As a result, the traditional shopping, recreation, travel, work, and socializing patterns are changing significantly. Physicality ceases to be a critical factor in defining a place: virtual settings may also have several qualities of physical places—they enable social interaction and present particular symbolic values for their users. A lack of tangibility does not prevent virtual places from gaining meaning for their visitors. Hence, the metaverse is generating a potential exodus of people from real places to digital ones (Oleksy et al. 2023).

Eventually place attachment increases one's desire to stay in a particular place and hinders the decision to migrate (Oleksy et al. 2023). However, the developing virtual worlds, offering excitement and safety, become a plausible alternative to physical places, provoking desire to migrate. In other words, virtual places are evaluated as good habitats (better than real places), and probably their users will consider migration to the metaverse.

For many reasons, migration to virtual reality is not equal to physical migration: the costs of moving are minimal, and migration decisions are easily reversible and usually have temporary nature, offering frequent relocation opportunities. On the other hand, virtual migrations and living in the metaverse become a truly spatial issue: they radically reduce the time spent within the real-world place of living (Oleksy et al. 2023).

The processes described in this section pose particular challenges to the everyday neighborhood existence as the place where people live and spend time. Who will meet friends in local public spaces, who will go shopping in the local grocery stores, who will go to the local barber shop if the residents prefer to move to the virtual world are the core question of the near future.

3.5 Servicization as a Trademark of the Twentieth Century

The digitalization effects are visible not only on the social level but also in the economy. As Toffler (1997) puts it, the epoch of agriculture was followed by intense transformations brought by the industrial revolution, which in turn was superseded by the era of information. Technological development caused a growth in service sector which gradually supplanted manufacturing and production. Intellectual labor displaced manual labor, changed lifestyles, and induced new forms of family life and new political conflicts (Fukuyama 2000).

The unprecedently high share of services in the economy marks a new stage in the development of humankind: since the mid-twentieth century, growing incomes have translated onto a growing demand for services. The so-called servicization is observable both in quantitative terms—as the number of economic entities related to services increases—and in qualitative terms, as the diversity of services grows (Damurski et al. 2019a, b).

Box 3.2: "Amazon Go" Stores

Amazon is one of the pioneer examples of "digitally native brands" which started opening their "brick-and-mortar" stores. Since the 2020s, this global shopping corporation introduced its hybrid service: Amazon Go store. The instruction is as follows: "Enter the store using your Amazon app or your credit card (…). Browse and shop just like you would at any other store. Anything you take from a shelf is automatically added to your virtual cart (…). When you're done shopping, you can just leave the store. A little later, your receipt will be available and we will charge your payment method." With 27 locations in the USA so far, it seems to be a very promising trend, offering a seamless shopping experience (Amazon 2022).

The first Amazon Go store, Downtown Seattle, USA. (Source: picture taken on November 2, 2019 by GoToVan and published on https://www.flickr.com/photos/47022937@N03/49005185111 under the Creative Commons license Attribution 2.0 Generic (CC BY 2.0). https://creativecommons.org/licenses/by/2.0/. Retrieved on 15/09/2023. No changes made. Origin: Fulbright Senior Award scholarship 2021)

Let me now try to list the factors determining the development of the service sector. Regardless of how we define a service—as a process of providing certain material or immaterial goods to individuals or communities (cf. (Kachniarz 2012)) or as a series of interactions between customers, processes, and material elements (cf. (Dominiak 2018))—there is a specific dynamics of this phenomenon. The range of services offered in a given territory changes in time, as a result of particular circumstances and conditions. The most important and influential factors are as follows:

- General economic situation of a particular area (GDP, efficiency in production sectors, condition of production and commerce, policies established and implemented by public authorities).
- Market demand (customers' needs and preferences, affluence of society).
- Supply aspects (provision of services, distribution patterns, prices, market and nonmarket divide of services, availability in various contexts).
- Social and demographic characteristics of a given community (lifestyle, free time, age structure situation, condition of families) (cf. (Flejterski et al. 2005)).

The crosscutting impact of servicization may be easily exemplified with the so-called liberalization of various sectors in the EU. In line with the neoliberal trends, European governments tried to improve the efficiency of public services by applying private sector management approach to public administration, to "run government like a business and make government work better and cost less" (Denhardt and Denhardt 2000, p. 550). As a result of the liberalization of service sector in the EU (which is a "general economic situation" factor listed above) since the late 1980s, many services have been gradually moved to open market competition (e.g., telecommunications; air, rail, maritime, and road transport; electricity and gas supply; eventually postal services).

It soon turned out however that private services work best in densely populated areas, where they reach high numbers of clients, but in some remote and less developed locations, service providers set higher prices or totally withdraw from the market (which demonstrates the impact of "supply factor" described above). Consequently, the liberalization process has diminished the capacity of nation-states to secure the accessibility, continuity, and affordability of public services (reflecting the "demand factor" specified above) (Colomb and Santinha 2014). Some arrangements introduced afterward in order to retain certain competences in the public domain, labeled as "social and territorial cohesion policies," have only limited impact on this situation.

The classical approach to servicization, based on evaluation of the economic situation, demand-supply relationships, as well as social and demographic factors, is being revised since the 1990s as a result of the vast implementation of the ICT in the sector. In particular, the geographical references of service nodes have been questioned, as the growing number of activities can be performed remotely. Therefore, in the following sections, the well-established spatial perspectives to services distribution will be elaborated, providing a background for further, critical discussion.

3.6 The Mechanism of Concentration and the Hierarchy of Services

Spatial distribution of services is driven by particular forces that exist in the market. Since the very beginning of urban civilization, the so-called concentration mechanism has shaped the land-use patterns in various scales. Services have been located in places where they could reach the maximum number of customers and where clients could minimize the need for traveling (Zipser 1983). To put things even more straight, this common rule reflects the principle of "least effort" (Zipf 1949) which states that all living organisms (but also well-designed machines) in any activity they undertake usually follow the path of the least resistance. This observation can be also made in relation to traditional urban services, clustering in particular spatial zones where the total operating profits exceed the total costs both for the customers and for the service providers.

The natural concentration mechanism described gains additional explanatory power in agglomeration economies which measure the benefits acquired when firms and people locate near one another. Clustering of services results in lower transport costs and easier exchange of goods, people, and ideas and thus greatly contributes to the success of large urban areas (Glaeser 2010). Moreover, agglomeration reduces transaction costs, fosters face-to-face contact, and stimulates the exchange of tacit knowledge. The specific patterns observed in human settlement network, comprising of nodes of various ranks, are a result of two overlapping factors: (i) the preference for higher population density near central places and (ii) the network of diverse centers offering various functions (Zipser 1983).

Probably the best known theory that describes the location patterns of services is Christaller's central places theory (Christaller 1933). Originating in the 1930s in southern Germany, it revealed that there is a hierarchy of services in human settlements, characterized with a hexagonal system with the biggest unit in the middle, some medium-sized cities located around it and high number of smaller units surrounding all bigger ones.

According to this theory, services can be allocated to several orders, ranging from low-order ones through intermediate ones to high-order ones. Customers travel a short distance to the low-order services which are of relatively low value and frequently used and require little skill to produce. As a result, there are usually many suppliers offering such services relatively close to the residential areas. In the intermediate order, clients potentially travel longer distances because those services are used less frequently and are of higher value. Similarly, very long distances are traveled to reach the high-order services that are highly specialized and used only sporadically by customers (Shearmur 2010).

This classification is easily translatable into a spatial logic: simple services are evenly dispersed in space, intermediate services are located in nodes with wider catchment areas, and high-order services tend to concentrate in one central place which offers a maximum accessibility to all potential clients in a large geographical zone. Hence, the centrality of goods and services offered in a city defines its position in the hierarchy of settlements and allocates an appropriate market area to it. An updated version of the central places theory presented by Lösch stated that settlements of the same size not necessarily offer the same functions and that larger places do not have to provide all the functions of the smaller places (Haggett 1965).

Following those concepts, Palomäki (1964) devised a hierarchy of centers which involved various kinds of services, such as administration, commerce, several public services, and recreation. The number of branches in a given spatial unit and the number of units in which a given branch occurs have been used to arrange spatial units on a graph, starting from the largest to the smallest ones. The contiguity of the graph was broken in particular areas, indicating the borderlines between the hierarchy levels in the settlement system.

Another interesting approach was developed in Poland in the 1980s as a "system of population servicing" (Nowakowski 1984). Despite the political context in which it was proposed, strongly influenced by the socialist ideology, it offered some interesting insights into how service nodes are created. Namely, there are two core criteria that shape the hierarchical network of services:

- The shortest travel time.
- The maximum attractiveness of the center.

The spatial distribution of lower-rank services, accessible within a walking distance, is determined by their maximum proximity to residential areas. The accessibility of higher-rank services is dictated by the availability of attractive functions and places, well-connected to the transportation network (Nowakowski 1984).

The hierarchy of service nodes can be observed at various scales, starting from the global, through continental and national levels, down to the local area. As a consequence, service centers in a city can be also divided to various types, like the huge multifunctional downtown (or central business district) offering specialized facilities to the citizens and incoming people, the middle-sized district centers, and the local service nodes in the neighborhoods satisfying the essential everyday needs of the residents. In line with the scope of this book, I will now focus on the last level.

3.7 Services in the Neighborhood

Small shops, schools, health centers, libraries, community centers, leisure centers, faith organizations, and transport stops form a network of neighborhood facilities where people cross paths and fulfil their daily needs. These centers of the local public realm are key neighborhood assets and building blocks of local identity (The Young Foundation 2010).

This simple approach to neighborhood services commonly presented by urban governance practitioners is a good starting point for a more academic, research-driven discussion on the range of local facilities. The core term here would be the neighborhood service center—a particular geographical location clustering various local economic and social activities.

This concept of the neighborhood service center is essentially connected with the hierarchy of urban nodes described in the previous section. As translated for the

American settlement context by the Urban Land Institute in the 1950s and reaffirmed over time, the Christallerian hierarchy may take a specific spatial and functional form of shopping centers. Such shopping center is a group of commercial establishments planned, developed, owned, and managed as a unit related in location, size, and type of shops to the trade area it serves. It provides on-site parking relating to the types and sizes of its stores (Urban Land Institute 2000).

There are five basic types of shopping centers, each distinctive of its own function: the convenience, the neighborhood, the community, the regional, and the superregional. In all cases, a shopping center's type and function are determined by its major tenants and the size of its area (Urban Land Institute 2000). For the studies presented in this book, three of them are most important.

Practitioner's Perspective
How would you define neighborhood services?

The range of services depends on the character of a neighborhood. In a residential neighborhood, stakeholders and the local community co-decide about the services they need in their area to provide desired quality of life. Neighborhood services can be divided into public ones and market ones. The first group includes public safety (police department, fire department), schools (but not childcare which due to the state regulations is within market sector), etc. *And market services include food provision (grocery stores offering healthy and affordable food—recently reintroduced to the Boston Downtown using zoning plans; restaurants), other small local shops.*

Ted Schwartzberg, Neighborhood Planner for Dorchester (north of Park Street), Department of Downtown & Neighborhood Planning, The Boston Planning & Development Agency [interview conducted via Google Meet 10/07/2021; reviewed and authorized by the interviewee on 13th of October 2021; origin: Fulbright Senior Award scholarship 2021]

Convenience retail, coffee shop, grocery, pharmacy, dry cleaner, banking, boutique-type retail, green spaces

Donald W. Birch, Executive Vice President/Chief Operating Officer Leggat McCall Properties LLC [interview conducted via MS Teams 9/09/2021 11:00 am-12:30 am; origin: Fulbright Senior Award scholarship 2021]

The first type is a convenience center with less than 30,000 sq. ft. (about 2700 m2) of gross leasable area, which contains a minimum of three stores. It is focused on personal services such as food, pharmacy, flowers, beauty and hair shops, as well as cleaners. Instead of being anchored by a supermarket, a convenience center is usually connected with a minimarket (Urban Land Institute 2000).

The second type is a neighborhood center which provides both convenience goods (foods, drugs, etc.) and personal services (laundry and dry cleaning, barbering, shoe repairing, etc.) for the surrounding neighborhood. It is built around a supermarket as the principal tenant and typically contains a gross leasable area of about 60,000 square feet (5600 m2) (Urban Land Institute 2000).

Box 3.3: Convenience Centers and Neighborhood Centers in the USA

A continuous research conducted in the USA by the Urban Land Institute in the 1980s and 1990s distinguished various types of service centers. Each of them was characterized by a particular mix of functions, represented by typical groups of tenants. The table juxtaposes two types of service centers relevant for neighborhoods: convenience center and neighborhood center (divided into metropolitan ones and small towns). The results suggest that neighborhood centers in small towns are the most multifunctional ones, commonly equipped with most of the listed types of services (Table 3.1).

Table 3.1 List of typical tenants in convenience centers and neighborhood centers in the USA in the 1990s (selection)

	Average number of stores		
Tenant type	Convenience centers	Neighborhood centers in metropoles	Neighborhood centers in small towns
General merchandise			
Variety store or discount store	0	0	0.4
Convenience market	0.1	0	0
Supermarket (full line)	0	0.4	0.6
Food service			
Restaurant	0.3	0.8	0.4
Sandwich shop	0.2	0	0.2
Pizza or Chinese fast food	0.4	0.5	0.3
Clothing and accessories			
Women's wear	0	0.5	0.4
Family wear	0	0	0.1
Drugs			
Drugstore/pharmacy	0.1	0.4	0.4
Personal services			
Beauty	0.4	0.5	0.3
Dry cleaner	0.3	0.6	0
Unisex hair	0.1	0.3	0
Nail salon	0.1	0	0
Medical and dental	0.4	0.6	0.2
Recreation/community			
Learning center/college	0	0	0.3
Financial			
Bank/finance company	0.1	0.3	0.2
Insurance	0.3	0.2	0

Source: Urban Land Institute (2000), (1997b), (1997a). Methodological note: the sample of "convenience" centers covered 80 areas located through the whole USA; in the "metropolitan" group, 20 US metropolitan areas are represented by 112 neighborhood centers; in the "small towns" category, data were collected from 107 shopping centers located in small cities, towns, villages, and rural locations outside metropolitan areas

Origin: Fulbright Senior Award scholarship 2021

Another level of the service hub in the ULI typology is a community center which—in addition to the convenience goods and services provided in the neighborhood center—offers a wider range of clothing (for men, women, and children), hardware, and appliances. A greater variety of sizes, styles, colors, and prices, followed by a higher specialty of stores, is a characteristic feature of the community center, even though it does not have a full-line department store. Its typical size is about 150,000 sq. ft. (13,900 m2) of gross leasable area (Urban Land Institute 2000).

This list of functions available at shopping centers in the twentieth century has to be complemented with the recent developments in the sector. As Pantano et al. (2021) note, modern shopping centers have recently undergone a transition from a functional profile focused purely on shopping to attractive locations for entertainment, dining, relaxation, and culture. Moreover, the development of multifunctional retail-leisure complexes has been stimulated and enforced by new technologies, offering customers an increasingly enjoyable experience (Pantano et al. 2021).

And even if those changes do not apply directly to convenience, neighborhood, or community centers, they indicate some kind of direction toward which the retail market is heading: multifunctionality. Bearing in mind the definitions presented above, a clear conclusion may be drawn: the hierarchy of service centers always starts at the local level. This confirms the importance of neighborhood governance which is co-responsible for local services and as such formulates the foundations of the whole urban settlement system.

For the purpose of this book, the term "neighborhood services" can be synthesized as the basic, essential amenities and facilities available or expected at the neighborhood level. Their presence (or absence), accessibility, and characteristics have a significant impact on the quality of life of both the local community and individual residents. The typical list of neighborhood services includes (but is not limited to) the following:

- Education (public kindergarten, elementary school, or similar services).
- Commerce (discount shop, bakery, grocery store, small market, etc.)
- Personal services (barber shop, beauty, laundry, etc.)
- Finance (ATM, small bank branch, etc.), healthcare (physician, dentist, drug store, etc.), food (cafe, restaurant).
- Culture (church, faith organization, community center, library, etc.)
- Administration (e.g., housing association, local authorities office).
- Transportation (bus stop or other public transport hub).
- Greenery (park or other open space) (Damurski 2020a, b).

All of those services are available at various spatial levels of urban areas, but at the neighborhood scale, they present a special value in shaping everyday human living environment. As such, they should be treated with appropriate attention.

Moreover, it is highly recommended that neighborhood services are spatially concentrated and organized in a certain manner as the local service center. This specific spatial distribution of neighborhood functions is a crucial condition of providing desired quality of life and local identity (Damurski 2020a, b). The concept of the local service center will be further discussed in the following chapter.

3.8 Suburbanization and Deurbanization Issues

The hierarchical models of the service centers discussed above provide a satisfactory representation of phenomena occurring in settlement systems typified by high-density housing. But a common phenomenon in the late twentieth and early twenty-first century has been suburbanization along with all its negative consequences: monofunctionality of large residential areas, low-density housing, poor communication with the city center, high level of car dependency, shortage of infrastructure and services (in particular commerce, recreation, and culture facilities), and a negative impact on the environment and landscape (Leboreiro Amaro 2014).

Moreover, in many parts of the world—be it America or Europe—suburbs encounter a so-called placelessness: the lack of meaningful spaces that would be used collectively and that would convey the specificity to a given location. Inhabiting places without history reduces social capital of the residents and impedes the formation of a community (Mantey 2015).

The universality of suburbanization stems from specific needs and aspirations of contemporary societies which—due to the conducive macroeconomic conditions and fashion—can be effectively met in the suburbs (to some extent at least). Nonetheless, the city remains essential as the place of work, education, and provision of services (Kajdanek 2012). The post-material needs focused around the quality of life, a sense of community, and self-fulfillment are typically realized by suburban residents outside of their place of residence—in the core cities (Mantey 2015).

In the recent decades, suburbanization has been accompanied by extensive dispersion of urban functions, which are increasingly flexible in terms of their location in space. Decentralization is in fact the "urbanization of the suburbs" (cf (Andersen et al. 2011)). It consists not so much in an outflow of residents but a shift of all types of activity (including services) from city centers to the urban outskirts (deurbanization). It seems that as long as a given activity can be transferred to a location with lower operating costs, the process of decentralization will continue (Hall 2002).

The dynamics of the urbanization processes in metropolitan areas in Poland confirms that suburban areas evince specific demographic and spatial transformations, including extensive development of residential and service functions (Kurek et al. 2013); (Maciejuk 2012). Some areas are privileged in this respect due to the availability of land, infrastructure, or simply fashion (Damurski et al. 2016); (Dąbkowski and Urbańska 2012), but some are not, which contributes to the growing polarization in urban development.

One thing appears to be certain: the decreasing density of urban tissue leads to a significant shortage in the availability of services. This happens in line with the natural concentration mechanisms described in one of the previous sections, which regulate supply and demand in a certain area. Even a gradual adjustment of the number of facilities to the growing population in suburban areas fails to solve the problem but rather contributes to further decentralization of urban functions in the larger scale (Damurski et al. 2016). Suburban residents are still forced to fulfil a proportion of their needs in core urban centers, which entails commuting to schools, shops, medical services, or cultural and entertainment amenities, generating a significant pressure on transportation systems. The overall result is an increasing spatial disorder, lack of public spaces, and deterioration of esthetic and functional values of the landscape (Wdowicka and Mierzejewska 2012).

In this context, a key task for local authorities in suburban areas is to ensure the inhabitants have access to services (public and commercial) at the desired level. This goal can be achieved by various policies and incentives, one of which is the concept of the neighborhood service center mentioned above.

3.9 The Problem of Shrinking Cities

Urbanization processes are typically described with quantitative data collected and presented per units of territorial administration. Tracing the dynamics of the population over longer periods of time allows identification of typical stages in the development of a city. The sequence is as follows: urbanization (where the number of residents of central area grows faster than that of the peripheries), suburbanization (where the number of the residents of peripheries grows faster than that of central area), deurbanization (where the entire metropolitan area loses residents), and finally re-urbanization (where again the population growth in central areas outpaces the one in the peripheries) (Lorens 2013).

Population count is thus a widely approved benchmark of the condition of a city. An urban area whose population is growing is considered "robust" and "healthy," since the influx of inhabitants is treated as a driving force of urban development, with an increasing demand for housing, proliferation of jobs and services, and a dynamic local economy. In turn, a decreasing number of residents is usually considered as a symptom of the decline of a city: people do not want to live there, construction sector slows down, unemployment is growing, and the city becomes less attractive to investors (Andersen et al. 2011).

In fact, in the beginning of the twenty-first century, 40% of large European cities (over 200,000 inhabitants) lost a substantial proportion of residents (Schlappa and Neill 2013). The geographical distribution of this phenomenon is not uniform, and the dynamics and patterns of demographic change differ between countries, but every state in the EU has shrinking cities in its boundaries.

Urban shrinkage is a result of broader processes, including globalization, leading to radical shifts in the economic base of a city, as well as structural processes related to production and consumption (cf. (Häußermann and Siebel 1988); (Martinez-Fernandez et al. 2012)). The result is always similar though: the decreasing number of residents entails lower revenues and higher operating costs; the maintenance of infrastructure is increasingly expensive; lower population density means a lower efficiency of public services; the sense of security is getting down; urban tissue loses cohesion as developed areas are separated by abandoned "no man's land." Lower levels of social and economic activity, and of the energy necessary for development in general, lead to the reduction in the number of businesses. A vicious circle emerges, bringing further threat of an intensified outflow of residents ((Rybczynski and Linneman 1999); (Martinez-Fernandez et al. 2012)).

Thus, urban shrinkage becomes a major challenge for urban governance. Reducing the political ambitions and adapting urban policies to the new size of population become a necessity (Leboreiro Amaro 2014): the growth-focused planning paradigm should be replaced with nongrowth governance, taking into account progressing demographic shrinkage (Martinez-Fernandez et al. 2012); (Adamski 2013); (Schlappa and Neill 2013). The depopulation of cities can be viewed as a unique chance for civilizational improvement—according to Kipta (2013), this is the first opportunity since World War II to replace quantitative development with qualitative growth by means of revitalization and other strategic policy solutions. Neighborhood services will have a special role to play in this process.

3.10 Availability, Accessibility, and Affordability of Services

Provision of services can be considered taking various perspectives—social, geographical, economic, or administrative (cf (Neutens et al. 2010))—each of which uses its own methods and provides different information. Such multiplicity of approaches may be synthetized with three terms—*availability*, *accessibility*, and *affordability*—describing particular aspects of the service provision in urban areas.

First, the concept of availability addresses the range and diversity of facilities offered in a given area. It involves both the typology of services (branches) and their spatial distribution in relation to other land uses (in particular residential areas).

Availability is modulated by the local market conditions: competition between the service providers on one hand and the customers' needs and preferences on the other. It is usually characterized by high dynamics, as some products or services may become unfashionable and simply disappear from the market, and some may be replaced with more advanced ones while some new products may be introduced.

The second term—accessibility—is seen as a key component in measuring the local quality of life and in ensuring a long-term economic, social, and environmental sustainability. It refers to public transport, different services, public spaces, and shopping centers (European Commission 2011). The meaning of this term is very similar to that of proximity described earlier in this book and is closely linked to travel between "sources" and "destinations" in the city. Accessibility is rarely considered as real because it is never based on full, comprehensive data (this would be difficult to do in practice). It is usually presented as a potential characteristic of a given area, based on the assumption that the destination of a journey can be reached in particular time under a given set of conditions. Accessibility studies produce maps of hypothetical time zones (isochrones) around certain facilities for particular means of transport (walking, cycling, car, public transport). Such maps can help to assess the number of residents within selected time zone and thus define the level of accessibility of different services—see, for example, the two-step floating catchment area method (Mao and Nekorchuk 2013).

Box 3.4: Services' Accessibility in Pereca Square, Wrocław

Accessibility analysis for Pereca Square neighborhood (Wrocław, Poland) was conducted in 2018 as a part of comparative research programme for five neighborhoods representing various settlement contexts (large urban cores, medium-sized towns, and suburban areas). This map was generated by ArcGIS module called Network Analyst using the "shortest path" procedure to designate catchment areas of particular services. The pedestrian network included all areas where walking was possible (sidewalks, park alleys, gates in buildings, squares, stairs, etc.), considering their various "impedance." Network Analyst returned polygons built by interpolation of linear geometry into a triangulated irregular network (TIN dataset). As a result, accessibility of various services and facilities was visualized in two time zones: 5 minutes and 10 minutes (Damurski 2021).

(continued)

Box 3.4 (continued)

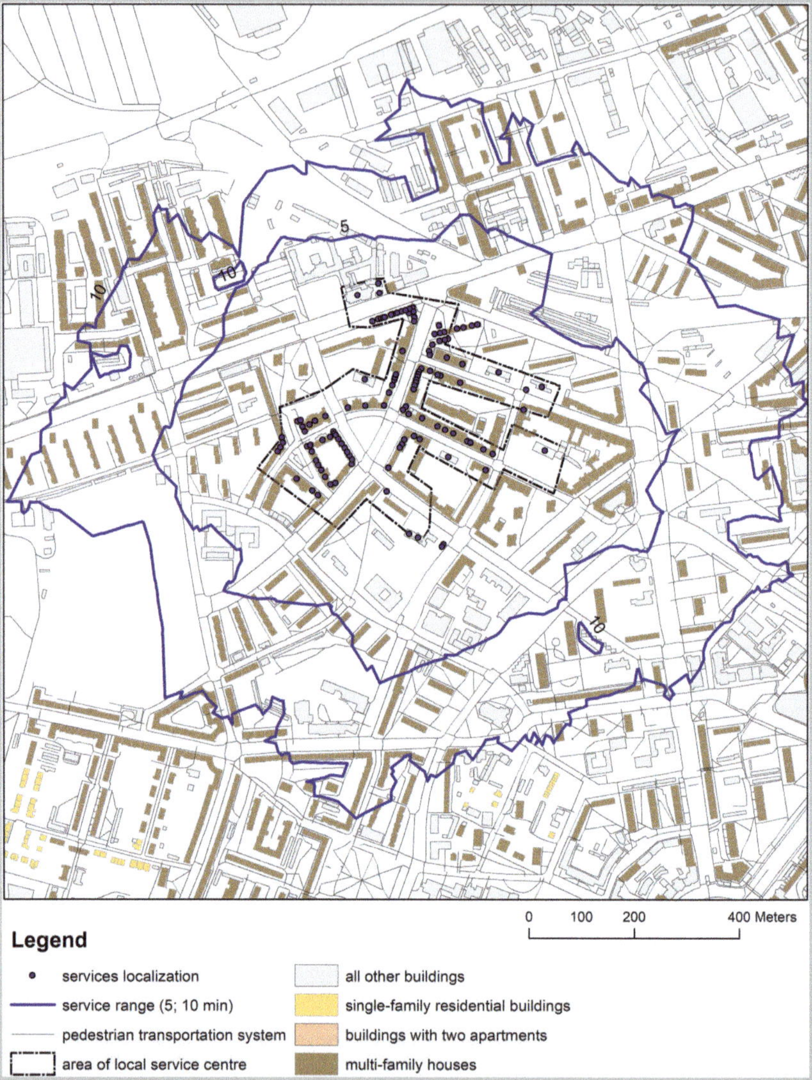

Compound accessibility map for various services in Pereca Square neighborhood (Wrocław, Poland). (Source: article by Łukasz Damurski published on https://link.springer.com/article/10.1007/s10708-021-10523-1 under the Creative Commons license Attribution 4.0 Attribution-NoDerivatives 4.0 International (CC BY-ND 4.0) https://creativecommons.org/licenses/by-nd/4.0/. Retrieved on 15/09/2023. No changes made. Origin: NCN research project 2016–2020)

Eventually the term affordability takes a more economic perspective and reflects the probability of using particular services by customers considering their buying power. It is expected to answer the question whether the services offered in a given location are not too expensive for the residents. Studies of this kind usually focus on a particular basket of essential products and services, comparing their prices year-to-year or in various locations.

However, the usefulness and explanatory power of availability, accessibility, and affordability as measures of urban services is being challenged by the ICT development. Great amount and great variety of online services are available to every internaut (unprecedented level of availability)—they can be accessed with one simple "click" (instant accessibility) and usually are cheaper than the offline ones (high affordability). Those distinctive features of the virtual service sector undermine the material structures of urban areas and require a revision of the to-date governance paradigms.

Adopting a neighborhood perspective, we need to recall one of the principle features of localism described in the previous chapter: rootedness and place attachment. Originally applied to local community members, it can be also adopted to the local services market. Recent changes in consumption practices include the shift toward more frequent shopping in small, local convenience stores and less interest in one-stop weekly shops in large shopping centers (Pantano et al. 2021).

For example, Polish citizens appreciate local businesses primarily for those of their qualities which are absent in large-format stores. Small local shops are better than shopping malls because they offer a personal relation with the customer (69%), convenient location (63%), and shorter queues (61%). The high quality of service (60%) and a satisfactory range of the goods in stock, some of which are not available in large chains (38%), are also mentioned (MAM Media 2018). Further, small shops have an important role to play in local communities, creating positive impact on local budget revenues, as well as purely interpersonal aspects.

Then the main threat to smaller retailers is the growing network of discount stores. But although the number of smaller shops is decreasing, sales per store are in fact constantly growing, and small- and medium-format stores remain the largest distribution channel of consumer goods in Poland (Polska Izba Handlu 2016).

Research on consumer habits and preferences in Poland unambiguously indicates a large and growing room for the development of traditional commerce. Forecasts of future consumer trends indicate that the prices in groceries are going lose importance as a factor shaping consumer choices, while convenience, a wide array of goods on offer, and a high quality of service are going to become increasingly important (Polska Izba Handlu 2016).

Those observations, even if limited to the Polish context, indicate that neighborhood services play an important, supportive role for localism. A diverse local market, comprising small retailers, offering easy access and personal relationship with customers, significantly contributes to the success of a neighborhood as a place of residence.

What about online services then? What is their impact on the local living environment? The next section commences this debate with a general overview of the novel developments in the twenty-first century.

3.11 New Developments in the Twenty-First Century

Bearing in mind various characteristics of the contemporary urban service sector described above, it can be generally said that the conditions in which services are being developed have been changing fast over the last years. The processes of metropolization and suburbanization, revitalization of city centers, and lifestyle changes in combination with technological developments have resulted in significant shift in the distribution of services, their quantitative and qualitative structure, as well as the manner in which they are provided.

The idea that various land uses cluster around central places, where people come together to engage in social and economic activities, becomes considerably more complex than ever before. The "localization benefits" derived from the concentration of manufacturing industries lose importance, while the richness and diversification of activities performed in cities and the quality of the urban environment become decisive factors in the location of firms (ECTP 2003). As a result, the multidimensional relations between the central hub and its hinterland are weakened (or even broken) and replaced with links with other metropoles, which are now the main partners in exchanging information, goods, funds, ideas, as well as investors or tourists (Lash and Urry 1994), (Frysztacki 1997). The well-known hierarchical structure of central places in human settlement is being ousted by the "space of flows" (Castells 1998).

Moreover, also the interactions within urban centers become more and more aspatial, nonmaterial. People no longer need to visit them in-person, because they can fulfil their demands in the virtual world. Consequently, the activities and land uses that define urban centers become essentially footloose, and we are unable to explain services location processes in such definite way as we were able in the past (Batty 2014).

An alternative and critical approach to the new developments of the twenty-first century points to the growing role of spatial proximity: despite the progressing virtualization of human activities, closeness of firms, workers, and clients still enables them daily functioning, including exchange of knowledge, distribution of goods, and provision of services (McKillop et al. 2009). According to this view, ICT cannot be perceived as a substitute of direct in-person interactions, but rather as a complementary channel of communication (Sztompka 2016).

Finally, another effect of the changing lifestyles and behavioral patterns is the increasing frequency of changes occurring in urban environment. While cities have always been evolving, mutating, and malleable, today uncertainty is becoming a kind of norm. The blurring of distinctions between land-use types empower temporality in urban spaces. So-called intermittent practices are planned or spontaneous activities materializing around a central purpose which is periodical and has a particular physical location (Allegri and Ochoa 2021). Intermittent practices put a significant pressure on existing urban structures and generate new challenges for policy, planning, and governance. The expected transformation of cities in response to this phenomenon should promote more flexible architectures, adaptable to the contemporary social and urban dynamics. Adaptability and responsiveness seem to be fundamental attributes in building sustainable, resilient urban areas (Allegri and Ochoa 2022).

In order to properly position the online-offline dichotomy in a wider civilizational context, we need to refer in detail to the characteristics of those two channels. An overview of the real and virtual aspects of contemporary services market is necessary to answer the question of territorial impact of the ICT in the neighborhood scale.

3.12 Online and Offline Channels

In the early 2000s, the retailing sector all over the world faced a challenge of unprecedented size: the birth of e-commerce. The emergence of Amazon, the disappearance of many shopping malls, and the expectation of many consumers that every product should be available within 24 hours at the doorstep are the significant signs of those times (Yoo et al. 2023).

More and more digital devices such as mobile phones and tablets, fueled with innovative online applications like social media, invade the traditional living environment. The multichannel (called also omni-channel) provision of services becomes a worldwide standard, and a typical consumer behavior follows a seamless shopping experience by using both online and offline channels in varying degrees. As a result, consumers' habits, triggered by advances in technology (e.g., mobile shopping) and the increased prominence of e-retail stores, are rapidly evolving into nonspatial ones (Pantano et al. 2021).

According to sociologists, two main patterns of using the Internet can be outlined. The first one relates to the users who treat online world an instrument supporting their work, school, shopping, and creative activities. The second one points to the people who use the Internet mainly as a source of entertainment (Batorski 2015). Online shopping is becoming increasingly popular among the more "instrumented" users, but a growing role of social media is also observed, supporting the creation of brand-consumer relationships and building mutual trust (Mills and Plangger 2015).

Box 3.5: Small Art Gallery Going Online

"Ładna sztuka"—a small art gallery located in historical part of Wrocław (Poland)—was one of the first visible examples of conversion to the Internet. Since the 1990s, it had been an atmospheric venue, selling antiques and non-professional art. However, in late 2018, it was closed and moved to the Internet, probably due to a high rent and low profitability of business (the premise was changed into a global brand café). An art gallery is not a typical neighborhood service, but "Ładna sztuka" example shows a broader trend in cultural service sector, strengthened significantly by the COVID-19 pandemic: in search for customers, cultural institutions organize events and promote their offer online.

The front door of "Ładna Sztuka" art gallery after its conversion to the Internet. (Source: Picture taken by the author in 2018. Origin: Author's own research)

To better understand those processes, we need to take a closer look at the online-offline relationships. But first, some conceptual clarification is necessary, as the lively virtualization process brings some new, sometimes confusing vocabulary like online services, e-services, digital services, virtual services, smart services, etc. In this book, I consider online services in general terms as all the market and nonmarket activities performed by service providers aiming at doing some specific work for customers using remote ICT. Consequently, an online service provider is the one who offers only such services; offline service provider is the one who does not run

Fig. 3.1 The information and fulfillment matrix. (Source: (Bell et al. 2014), p. 47. Permission for reprint obtained from the copyright owner on 11/09/2023)

such services at all; and eventually multichannel (or omni-channel) service provider is the one who operates both online and offline services, using the two channels for communication and distribution of his or her services and products.

In an multichannel retail environment, customers can either visit stores to obtain information, or they can seek information remotely. They can also either visit a store to pick up items, or the store can "come to them" when products are delivered (Fig. 3.1).

While all retailers need to manage fulfillment and information effectively, there are important nuances to how this is done. From the consumer's point of view, there are both advantages and disadvantages to picking up a product while visiting a physical store. The client does not have to pay for shipping or wait for the product to be delivered; offline shopping is also rated higher for the enjoyment of the shopping experience, the ability to see-touch-handle the product, personal service, hassle-free exchanges, and fast delivery (Levin et al. 2003). However, a consumer of a physical store incurs travel costs (Bell et al. 2014).

The advantages of the online channel include the immediate and comprehensive display of information and the ease of comparing the features and prices of particular products. Consumers perceive online stores as better for shopping quickly, with a large selection and access to more products with a wider range of features. Online shopping is also perceived as the source of the best prices (Levin et al. 2003). Benefits also include reducing travel costs and having access to products that would not necessarily be displayed in a physical store (Bell et al. 2014). On the other hand, lack of personal service, inability to inspect or handle the product, and concerns about delivery and payment are perceived as disadvantages of online shopping (Levin et al. 2003). Waiting time (and the resulting delayed gratification) and shipping costs can also be perceived as unfavorable compared to offline stores.

From the retailer's point of view, the challenges of fulfilling orders in-store or via home delivery are very different. If orders are fulfilled in physical locations, there are important decisions to be made about store design, accessibility to customers, and appropriate size for stocking. In addition, fulfilling in-store transactions requires the retailer to have the right products in the right stores at the right time, which involves making informed decisions about which products to carry in each physical location (Bell et al. 2014).

Conversely, fulfilling orders via delivery relaxes some of the constraints for the retailer's physical locations. First, fulfillment can be centralized and performed from a distribution center located in a less expensive area. Second, a central delivery management simplifies managing the items in stock because it allows forecasting at a more aggregate level, reducing supply-demand mismatch costs (Bell et al. 2014).

New challenges are now generated by the emergence of the metaverse. This online environment (described earlier in this chapter) is expected to fundamentally change how consumers interact with the digital world and reshape the commercial

Box 3.6: Potential for Offline-Online Conversion

A comparative analysis of five neighborhoods, representing various settlement contexts in Poland (large urban cores, medium-sized towns, and suburban areas), was conducted in 2017 within the research project "Local service centre as a tool for enhancing territorial cohesion of urban areas" (grant number 2015/19/B/HS4/01301). Below two of them are juxtaposed to show the share of online services users and nonusers among customers and service providers. In Pereca Square (Wrocław), a neighborhood located in a core urban area, the potential for conversion among customers is relatively higher than in Rynek (Siechnice), a suburban neighborhood. This suggests that residents in peripheral urban zones are more willing to switch to online services, probably in order to reduce the need for traveling. However, it is the opposite situation regarding service providers: conversion potential is higher in the suburbs than in the urban core. This may mean that service sector in peripheral areas is still more traditional, operating mainly in in-person mode (Table 3.2).

Table 3.2 Potential for offline-online conversion in Pereca Square (Wrocław) and Rynek (Siechnice)

Group of respondents	Neighborhood					
	Wrocław, Pereca Square			Siechnice, Rynek		
	Nonusers	Users	Potential	Nonusers	Users	Potential
Customers	86 (41.0%)	121 (57.6%)	0.7	18 (25.4%)	53 (74.6%)	0.3
Service providers	43 (74.1%)	15 (25.9%)	2.9	13 (81.3%)	3 (18.8%)	4.3

Source: (Damurski et al. 2018)

Origin: NCN research project 2016–2020

landscape. Retailing in collaborative virtual setting allows consumers to navigate through an immersive virtual space using their avatars and to interact with other avatars of retail personnel and other consumers. Like in e-commerce, retailers who successfully establish a metaverse presence can expand their reach, enabling consumers from distant locations to visit virtually their stores (Yoo et al. 2023).

Although the metaverse is still in its infancy, the opportunities for retailers to grow and develop their brands in the digital world seem endless. The metaverse offers unique opportunities at all stages of the customer journey, including prepurchase, purchase, and post-purchase, complemented by economic exchange, social relationships, and interaction with the environment.

However, with these benefits come certain risks, including the difficulty of transferring offline products directly into the metaverse (prepurchase stage), providing in-person inspection (purchase stage), and offering limited upgrade opportunities to customers (post-purchase stage). As the metaverse continues to shape, expand, and evolve, retailers are constantly exploring and optimizing how they interact with consumers. In the long run, they will undoubtedly find relevant solutions to manage the risks and effectively address the growing needs of customers in the metaverse (Yoo et al. 2023).

Considering all the aspects discussed above, it may be forecasted that the best sellers will win the omni-channel revolution by working across the permeable boundaries of information and fulfillment, offering the right combination of online and offline experiences for the customers, taking advantage of the new functionalities of the metaverse. In the next section, several scenarios of this situation are considered.

3.13 Four Multichannel Options

This section deals with particular forms of virtualization of services: offline retailers getting online-only, offline stores introducing online channels, online retailers opening their offline outlets, and ultrafast delivery of everyday products. Each of them demonstrates specific aspects of the online-offline relationship and may serve as a reference for further development of urban policy options.

1. Offline retailers getting online-only. In some extreme cases, the total transition of traditional services to the virtual environment may bring long-term changes to the structure of the market, referred to in the literature as a slow burn (cf (Pendall et al. 2010)). For example, bank branch closures in the UK in the last decade as a result of online conversion of financial services brought severe shortage in accessibility of banking in peripheral areas. This process impacted socially vulnerable groups such as the elderly and those on low incomes who are likely to be most dependent on physical banking facilities such as cash deposits or withdrawal. Also, some community-wide economic impacts on the sustainability of

local businesses and the state of local high streets have been observed (Langford et al. 2020). Therefore, this solution is the least favorable from the neighborhood perspective.

2. Offline stores introducing online channels. Modern retail marketing increasingly utilizes new technology-based services that merge the advantages of physical stores with those of online shopping. The wide spectrum of traditional factors influencing consumers' choices of shopping centers (such as the layout, being a nice place to spend time, quality of the stores, transportation, distance, availability of parking, and so on) is now complemented by new elements. These include (i) the promotion of green initiatives; (ii) the ability to provide memorable experiences; (iii) the presence of particular areas and places like parks, lake views, etc.; (iv) the availability of additional functions like clubs, bars, and food court; (v) the utilization of particular technologies such as mobile apps and AI; and (vi) social media channels to directly interact with shopping center managers and other consumers (i.e., finding the promotions and events, delivering personalized messages, and digitally embracing consumers while in the store) (Pantano et al. 2021).
3. Online retailers opening their offline outlets. This process initially involved showrooming (when customers can inspect products in the physical store and then buy them online) and webrooming (when customers research products online but purchase them in a physical store) (Kumar et al. 2017). Research has shown that webrooming has a positive impact on customer satisfaction and that showrooming is related not only to price savings but also to other factors such as perceived quality gains. As a result, greater accessibility to both distribution channels contributes to assuring the customer's maximum satisfaction with their purchase (Cortiñas et al. 2019).

From the retailer's point of view, the expansion of online-first retailers into offline stores serves the purpose of "supercharging" customer value. Many of the digitally native brands are transforming their outlets from fulfillment-dominant centers into experience-dominant centers, improving the customer satisfaction, attachment, and loyalty. Such stores offer the retailers the opportunity to observe and characterize customers who enter their premises, including their nondigital attributes such as emotional and sensory response to products or the interior design (Bell et al. 2018).

But also the customers benefit from offline stores, where they can resolve potential uncertainties about the attributes of the products. Showcasing products in showroom-like areas activates customers' creativity as they can imagine how the products would look in their homes. That is another reason why showrooms have changed into experience centers where customers can engage meaningfully with the retailer and with the product (Bell et al. 2018).

4. Ultrafast delivery of everyday products. The so-called quick commerce solutions refer to on-demand delivery of items from multiple categories in under 1 hour. After a customer places an order on an app or web-based platform, a quick commerce operator picks up the order from a retail location and delivers it to the

customer. Q-commerce focuses on small quantities of grocery and convenience items easy to delivery within a relatively short time. In March 2020, amidst the first wave of COVID-19 lockdowns, orders placed in Glovo—one of the leading players in this market—more than doubled from the month prior (Moreno and Barnett 2021).

Today, an advanced business model—ultrafast delivery—is being introduced. Following the same rules as Q-commerce, it is additionally supported by solutions like "dark store." Dark stores are warehouses located in densely populated urban areas, within a distance of approximately 2 kilometers from their potential customers. They are approximately 300–600 square meters and stock commonly ordered products, and as a rule, they are not open to customers (though some locations offer a customer pickup option). Dark kitchens develop in a similar way, offering instant delivery of food prepared in a closed manufacturing site located close to the customers. The delivery time in this model drops down to as little as 10 minutes (Moreno and Barnett 2021).

The trend of ultrafast delivery is to some extent halted by the "last mile" effect, which is probably the biggest obstacle to the success of online grocery shopping. Inefficient deliveries to private customer households in more distant urban areas lead to substantial increase in the cost of such service. Several solutions dealing with the "last mile" issue, based on technological developments and innovations, are currently being developed all over the world. The most promising include drones, autonomous delivery robots, crowdshipping, and public transport (Boysen et al. 2021).

The four options of multichannel service provision presented in this section are probably not the only ones observable in the market. However, they offer a good insight into the range and variety of activities undertaken by the service providers in building their omni-channel performance. In the following section, the consequences of online conversion and hybridization will be discussed, as a background for approaching urban governance challenges related to virtualization.

3.14 Conversion, Hybridization, and Their Consequences

As it has already been stated in the previous section, material (on-site) forms of providing services are now complemented by virtual (remote) ones. Depending on the relationship between offline and online channels, this process is referred to as conversion (when the whole facility is transferred to the Internet) or hybridization (when some components of a facility are offered online but others remain available offline) (Damurski et al. 2019a, b).

Box 3.7: Ultrafast Delivery

Rising customers' expectations regarding the time of delivery and the range of products delivered induce new solutions in logistics. This applies especially to food and grocery items, which require ultrafast delivery due to very short expiry date. A growing number of companies such as Glovo, UberEats, Troav, DoorDash, Pyszne.pl, or Lisek offer on-demand, door-to-door delivery services, using mostly bikes and mopeds. This process fosters a profound change in consumer behavior (reduced need to travel to shops and restaurants) and affects the functioning of many services (focusing on serving couriers instead of walk-in clients).

Examples of ultrafast delivery modes in Boston and Wrocław. (Source: Pictures taken by the author in 2021 and 2022. Origin: Author's own research)

The introduction of online channels in the service sector has challenged the foundations of the entire market, affecting both individual companies and global networks (Lieber and Syverson 2012). The potential of a particular retailer to become a multichannel one depends on the advantages and disadvantages perceived by consumers and by the retailer itself. We have already addressed this issue in the section dedicated to online and offline channels: in order to develop an optimal mix of online and offline functionalities, service providers need to determine which components of their products are better delivered face-to-face and online (Levin et al. 2005).

At this point, we could also mention the division between market and nonmarket services (the latter also known as services of general interest). However, this dichotomy is not really relevant in this case as the whole economy is becoming virtualized to some extent and both market and nonmarket providers are introducing new forms of delivery. Terms such as e-commerce on the one hand and e-government, e-administration, e-healthcare, e-learning, and many other "e-'s" on the other reflect the changes taking place in both sectors. Despite some differences in their content and nature, they share similar features in terms of the establishment, management, maintenance, and development of services. Therefore, market and nonmarket services do not require a separate approach in the research presented.

In multichannel mode, services are delivered simultaneously through conventional and online environments that are complementary and not interchangeable. When face-to-face and computer-mediated channels coexist, they have a mutually positive effect on service quality (Wang et al. 2016) (Piercy and Archer-Brown 2014).

However, consumer preferences for online and offline services differ for different products at different stages of the service delivery process. "High-touch" products are those that require consumer inspection at each stage of the shopping experience (e.g., vegetables). In contrast, "low-touch" products are those that do not require inspection to assess their quality before purchase (e.g., airline tickets) (Lynch et al. 2001). Other products fall at different points on the continuum (Damurski, et al., 2019).

Practitioner's Perspective
What is the advantage of traditional retail?

People want to experience products before they buy them.

Donald W. Birch, Executive Vice President/Chief Operating Officer Leggat McCall Properties LLC [interview conducted via MS Teams 9/09/2021 11:00 am-12:30 am; origin: Fulbright Senior Award scholarship 2021]

A study conducted in 2003, when the era of online services was just beginning (Shankar et al. 2003), found that while customer satisfaction with an online service was the same as with an offline service, loyalty to the service provider was higher in the online channel. Ten years later, this trend became even stronger—measures of customer satisfaction and willingness to recommend a particular store to others were five times higher for online shopping than for offline commerce (Nowicki and Wojnarowicz 2015). At the same time, the relationship between service quality and customer loyalty was similar for online and offline channels (Walsh et al. 2010). In other words, satisfaction was the determining factor for customer loyalty in both physical and virtual stores. This proves that the two channels have a lot in common in terms of marketing performance.

Box 3.8: Biofamily: Post-pandemic Conversion

Biofamily is a grocery specialized in ecological food and cosmetics. It was established in Wrocław in 2021 as a brick-and-mortar supermarket offering contactless 24/7 access for mobile app users. This model worked perfectly during the COVID-19 pandemic when many restrictions came to force, reducing the need for in-person meetings. However, after the pandemic, in early 2022, the store was closed and moved to the online-only format, probably due to significant competition in ecological food market and a global economic crisis.

Biofamily shop exposition after conversion to the Internet. (Source: Picture taken by the author in May 2022. Origin: Author's own research)

Practitioner's Perspective
Digitally native brands opening their offline stores

We are finding more interest from the "digitally native brands" (like Warby Parker or even Amazon). They started online, but they have found that after opening their brick-and-mortar stores in particular geographical areas, their sales have increased (...). Because they are there, there is brand recognition (so it is marketing), and someone may order something online and have it fixed or augmented in the store. That trend is expanding. Haired with some urban planning, architectural answers, the pickup zones in those urban locations most of the online retailers are changing their model.

Patrick McMahon, Senior Vice President, Regional Development, Federal Realty Investment Trust, Somerville [interview conducted via Zoom 12/06/2021; reviewed and authorized by the interviewee on 11/19/2021; origin: Fulbright Senior Award scholarship 2021]

It is symptomatic that neighborhood services are featured by relatively low online conversion indicators compared to citywide sectors like clothes, books, or electronics. This may mean that the local market is still managed with traditional offline retail channel. In 2014, in Poland, only 12% of online customers bought food online (Plesko and Świderski 2015), although the food sector is currently massively expanding its potential. As Shearmur (2010, p. 46) notes, "even groceries can now be ordered online: the local grocer no longer needs to be local!". This observation confirms the general trend of overwhelming virtualization.

The recent COVID-19 pandemic and the resulting lockdowns brought an unprecedented impulse for the online sector expansion: most services have adopted multichannel solutions which was necessary to survive that difficult time. Today online and offline shopping channels are so deeply interdependent that it is hard to distinguish them. An average customer can start his/her consumer journey wherever on the Earth, at any time using his/her mobile device, searching online for the products he/she needs. Then he/she can travel to the offline shop using mobile mapping service, try the product personally, and buy it or not. Alternatively, the customer can order the product online, with a door-to-door delivery, try it at home, and decide whether to keep it or to return it the same way. Satisfaction and loyalty are shaped by multiple factors which cannot be divided into online and offline ones anymore.

Box 3.9: Hybridization of Services in Boston

In 2021, a pilot street survey was conducted among service providers in two neighborhoods in Boston metropolitan area. These were Fields Corner in Dorchester (a well-established neighborhood with relatively long history and a naturally evolving system of service centers located around the main streets) and Assembly Square in Somerville (a dynamically developing modern neighborhood with booming community and vibrant European-style public spaces with shops and restaurants). The results suggest that hybridization of services proceeds much faster in newer neighborhoods, where competitive market fosters innovation, than in the older ones, where local economy is still based on traditional, in-person interaction.

Charts showing the results of a pilot survey with service providers. (Source: Author's own research. Origin: Fulbright Senior Award scholarship 2021)

3.15 Apparent Online-Offline Dichotomy

Then what is actually the difference between traditional and online services? Four main criteria of differentiation can be considered here (cf (Damurski, et al., 2019)):

1. Space. Each service provided in a traditional way requires a specific place enabling a direct contact between the customer and the retailed (e.g., a shop, an office, a studio). Online services reduce the role of spatial location of the service

facility—most products can be delivered anywhere, if only the customers have access to appropriate transmission networks.

2. Time. Traditional services are usually open at certain hours of working days. Therefore, an important advantage of online services is that they can be ordered by customers at any time and that they can be performed by service providers at a convenient time (e.g., at weekends or at night). Moreover, some services can be performed instantly by automated online systems (e.g., buying tickets or audio files).
3. Information. Originally, customers searching for services could only consider offers which they knew through personal experience or peer recommendation and that were closest to their location. This was one of the reasons why the concentration mechanism came into force, based on the limited availability of services within a reachable area. Today, customers have access to up-to-date information on services all around the world. They can search for and buy any offer that meets their needs and expectations. This requires specific competences both from the customers (finding reliable information in the complex and overloaded databases) and services suppliers (providing comprehensive and searchable information about their products), but this cost is always lower than searching through traditional services and advertising them in physical space.
4. Social interaction. Aspatiality of online services have certain social consequences: the lack of face-to-face contact with the customer. The result is a progressive anonymization of interactions in the service sector, contributing to a wider trend of depersonalization of human relations. People focus on satisfying individual needs and the Internet facilitates that. Interpersonal relationships become increasingly superficial, which is in the contrary to the localism trend with its desire for personal relationships between the customers and retailers in the neighborhood market.

Despite the differences described above, the way of providing online services paradoxically is not really different from the one we have known for centuries. Above all, the remote provision of services is in line with the least effort principle (cf. (Zipf 1949)): the Internet significantly reduces the cost and time needed to use a service. In addition, as with offline services, the goal is also the same: achieving an optimal price/quality ratio, bringing customer satisfaction, and inspiring loyalty to the service provider (Schmidt 2015).

Practitioner's Perspective
Online versus offline services

There is no online versus online distinction, but there are different things that customers do online and different things that customers do offline.

Antonio Moreno, Sicupira Family Associate Professor of Business Administration, Harvard Business School; Unit: Technology and Operations Management [interview conducted via Zoom 11/04/2021; reviewed and authorized by the interviewee on 14th of November 2021; origin: Fulbright Senior Award scholarship 2021]

Simultaneously, significant changes take place in human territoriality. As it was mentioned earlier in this book, the Internet followed by the metaverse offers user experiences which can successfully enhance (or sometimes even replace) the traditional, place-based behaviors. This is another argument against a strict division between online and offline services: the contemporary local market should be perceived and managed as a real-virtual continuum.

All of those observations suggest that the online-offline dichotomy is apparent and purely abstract—it does not affect the essence of the service provision process. And here a central question arises: how does this vision correspond with the traditional, hierarchical network of service centers? Are we witnessing a fundamental shift in human settlement patterns or just an evolutionary adaptation to new technologies?

Box 3.10: Madison Reed Offline Location Strategy

Madison Reed is a US hair color company whose products are crafted in Italy according to strict European Union (EU) safety standards. The firm exists mainly online but has also a growing network of offline stores all over the USA (86 locations in 2023). The Madison Reed Hair Color Bar in Somerville, MA, offers a selection of personal services (various hair coloring options), free color consultation, and retail of company's products. Clients can book their visit online through a countrywide website. An employee interviewed in 2021 described a specific development strategy adopted by the company owners in searching for new locations of their offline outlets: they track the online customers' credit card addresses, and they open new services in locations with the highest concentration of clients. As a result, their Hair Color Bars serve mainly the local clients, living in particular neighborhoods (Madison Reed 2023).

Box 3.10 (continued)

Madison Reed Store, Flower Hill, NY. (Source: Picture taken by AITFFan1 on 30.07.2023 published on https://commons.wikimedia.org/wiki/File:Madison_Reed_Store,_Flower_Hill,_NY_July_30,_2023.jpg under the Creative Commons license CC0 1.0 Universal (CC0 1.0) Public Domain Dedication https://creativecommons.org/publicdomain/zero/1.0/. Retrieved on 15/09/2023. No changes made. Origin: Fulbright Senior Award scholarship 2021

3.16 Agglomeration Effects Unthreatened?

As it was already stated in this chapter, agglomeration economies measure the benefits that come when firms and people locate near one another. This is why the hierarchy of human settlements emerged in the history of civilizational development.

However, in the late twentieth century, as transportation and communication costs decreased, spatial concentration and agglomeration mechanisms seemed to become less important. Urban land uses have loosened their requirements regarding geographical positioning and became more flexible in terms of location. Those structural changes in the service sector have been ambivalently referred in urban geography literature. Some authors claimed that human interactions within cities become more and more virtual, transferred to the metaverse, and that urban activities become essentially footloose. According to this view, the future of urban areas is uncertain because we are unable to explain location phenomena in the definite way we were able to in the past (Batty 2014).

Practitioner's Perspective
What is the impact of the Internet on urban areas?

Depending on the character of the center, the Internet brings development (like in large department stores where ordering products online and picking them up in store became one of the ways of shopping) or decline (like in small centers where some retailers lose their customers to competitors based online which results in vacancies). Therefore, online shopping has a real impact on land use.

Ted Schwartzberg, Neighborhood Planner for Dorchester (north of Park Street), Department of Downtown & Neighborhood Planning, The Boston Planning & Development Agency.

[interview conducted via Google Meet 10/07/2021; reviewed and authorized by the interviewee on 13th of October 2021; origin: Fulbright Senior Award scholarship 2021].

The more conventional stores don't need as much square footage as they ramp up their online platform and their ability to find their customer online. They still want a brick-and-mortar, but they don't need as much as they used to.

Patrick McMahon, Senior Vice President, Regional Development, Federal Realty Investment Trust, Somerville [interview conducted via Zoom 12/06/2021; reviewed and authorized by the interviewee on 11/19/2021; origin: Fulbright Senior Award scholarship 2021]

This argumentation, however, can be contrasted with the concept of "e-topia" (Mitchell 2000); (Mitchell 2005) describing the spatial consequences of the ICT development. This theory states that we cannot treat virtualization as a threat to urbanization. First, the investments in telecommunication contribute to the diversity of urban areas by merging various functions in one place (a building, a neighborhood, a district), including home and work. Second, the widespread wireless standard in the ICT makes actually every space (a hotel room, a table in a restaurant, a bench in a park) a potential workplace. Even if the current developments decrease the demand for office space, they simultaneously remain substantially urban in their very nature, enabling human interaction, and stimulate creativity, cultural activity, and technological progress.

This approach is supported by other authors who suggest that agglomeration effects are unthreatened as the growing usage of ICT leads to an increased need for spatial proximity (I will elaborate on this further below). Consequently, urban areas provide unbeatable localization values for innovation and growth (Shearmur 2010).

This observation is not novel though. Already in the 1990s, Lusht and Farber (1996) predicted that new communication technologies will result in both centralization and dispersion of human settlements. A reduced need for spatial accessibility of some functions (such as production or distribution) will be accompanied by a rising need for spatial concentration of other functions (including decision-making, management, and control), which require a direct face-to-face interaction.

Box 3.11: mBank Launching its Offline Channel

mBank was established in 2000 as the first exclusively online retail bank in Poland. As a result of rebranding in 2013, it became a combination of traditional banking offering comprehensive services in financial service centers throughout the country with banking available via the Internet and by phone. Having 143 retail branches and 166 mKiosks throughout Poland in 2019, in both large city centers and low-populated and economically disadvantaged areas, it started to compete with the existing traditional banking sector. However, the results are scarce so far: its offline locations seem to have small number of clients compared to other well-established banking brands in the country.

mBank retail branch in one of the shopping malls in Wrocław, Poland. (Source: Picture taken by the author in August 2022. Origin: By the courtesy of Marta Cząstkiewicz)

The history of civilization shows that every invention which has reduced the cost of transport has likewise contributed to dispersion of human settlement (Lusht and Farber 1996). This is also the case today due to the rapid ICT development, even though the dynamics of this process makes it hard to evaluate its impact on settlement patterns. On one hand, the lower cost of communication and transportation is weakening the concentration mechanisms and reducing agglomeration effects. On the other hand—paradoxically—despite the ease of flow of goods and knowledge, urban centers remain stable and lively (cf (Glaeser 2010)).

Moreover, information and communications technologies and face-to-face contact rather complement each other than substitute themselves. The growing quantity, variety, and complexity of the information produced leads to an increased need for spatial proximity which facilitates face-to-face contact, enhances information exchange, and encourages learning and innovation in certain local contexts (McCann 2003) (Shearmur 2010). Yet, a central paradox of our time is that urban agglomerations remain remarkably vital, despite ever easier movement of goods and knowledge across space (Glaeser 2010). Even if they are characterized by nonlinearities, they point to the advantages of physical proximity: a short distance to other firms, workers and consumers may help businesses in their effective day-to-day performance (McKillop et al. 2009).

Finally, a balanced approach says that due to telematic revolution, contemporary urban geography contains both a dynamic of dispersal and of centralization (Sassen 2005). The fact that it is not necessarily clients who travel to service providers but the providers who travel to clients does not affect Christaller's (1933) statements on the settlement hierarchy, although it does have repercussions on how service sectors function (Shearmur 2010).

Thus, the agglomeration effects described above seem to be unthreatened, at least when referred to whole urban areas. When observed at the neighborhood level, the situation seems to be slightly different though: the scale of the local market is much smaller, and the mix of land uses is much less diverse than in the citywide scale which makes it difficult to apply the same measures and rules. And this is one of the reasons for writing this book.

3.17 Intermediate Conclusions

This chapter presented various developments in the ICT sector in the recent decades as well as civilizational contexts in which its impact may be observed. Immersive virtualization of everyday life brings profound consequences to human behavior and as a result to the performance of urban areas.

In particular, the processes of online conversion and hybridization of urban services have been analyzed, pointing at the fact that there are currently two main kinds of items (goods and services) on stock: experiential and recurrent ones. The first one requires a direct, in-person interaction and inspection of the product, while the latter one can be performed totally online.

This division raises particular issues in service provision in urban areas. As customers become "omni-channel" in their thinking and behavior, sellers need to be as well; online functionalities initially perceived as "nice add-ons" are becoming "must-haves." The question for service providers is no longer whether to operate an omni-channel strategy but how to implement it most effectively (Bell et al. 2014).

As a result, most of the services become to some extent multichannel activities, following various paths of development. Each of the available options of online conversion and hybridization generates particular challenges to local urban governance:

- Offline retailers getting online-only reduce the accessibility of services in a particular area which can have significant, negative impact on particular social groups.

- Offline stores introducing online channels are becoming multifunctional experiential centers with much wider land-use program and a much more diverse customer profile.
- Online retailers opening their offline outlets introduce new kinds of services (interactive showrooms) of relatively small size that can generate added values in local public spaces.
- Ultrafast delivery of everyday products results in the emergence of dark stores and dark kitchens located in the neighborhoods, generating courier traffic but no direct customer-retailer contact which bears specific consequences to the surrounding areas.

The paradox of our times is that "offline is dead and dying, yet it is also alive and thriving" (Bell et al. 2018, p. 1). Despite the visible threats of virtualization, urban services are very much alive, focusing on effective delivery and shifting from pure fulfillment to experience-oriented environments. Smaller in size, multichannel, tech-enabled, and high-touch spaces are becoming the norm.

What is noteworthy is that online conversion and hybridization of services do not change their essence nor core characteristics. It is still the demand-supply relation that shapes the market, aiming at satisfying customers' needs and providing relevant revenue to the service providers. However, the way of delivering particular goods, products, or solutions has changed, altering the urban land-use structures.

Aspatiality as the main feature of online services reshapes the everyday practices of urban residents, reduces the need for traveling, modifies the style of social interactions, and challenges the classical concentration patterns. It requires a greater caution from the urban governance practitioners. In some cases (in particular urban cores), agglomeration effects remain stable and valid and need only some minor adaptation of the policy goals. But in other places (especially neighborhoods), it calls for a revision of paradigms and a shift in governance patterns.

Therefore, after considering various aspects of online conversion and hybridization of urban service sector, it becomes clear that contemporary local urban governance has to respond to the virtualization processes with appropriate policies and practices in order to support the core neighborhood functions and structures. In the following chapter, I will demonstrate how this can be done and what capacities are necessary for building a multichannel neighborhood governance.

Chapter 4
Governance as a Multichannel Process

Abstract This chapter describes the recent changes in the urban governance related to the ICT development. Drawing on the overview of virtualization phenomena presented in the previous chapter, several aspects of public decision-making processes are discussed. The first sections describe two models of public administration (traditional and collaborative), outline the core characteristics of public communication, define the concept of urban discourse, and point to the formal and informal elements of the decision-making processes. Then, in the second part of the chapter, various policy approaches adopting online solutions are discussed: e-administration, e-governance, e-planning, e-participation, and others are presented. The final sections introduce some multichannel solutions to the neighborhood services sector, which leads to the conclusions of the whole chapter.

Keywords Urban governance · Citizen participation · E-participation · Urban discourse · GIS · E-planning · E-governance · Local service center

4.1 Two Models of Urban Governance

Public policy and administration in the early twenty-first century are increasingly challenged by complexity, fragmentation, uncertainty, and multiculturalism (Belof 2013). Traditional democratic procedures prove to be inefficient in fulfilling the demands of contemporary urban areas. As a result, alternative solutions are being developed allowing societies and communities to participate in decision-making.

This book started with a critical review of urban governance paradigms. One of the main, repeatable components in most of the presented approaches was a postulate to enforce democratic standards in decision-making, balance all relevant interests, and provide citizen participation, communication, and monitoring. Before starting the discussion on how this can be done using multichannel solutions, it is worth to demonstrate the background of the current, participatory principles in urban governance.

Ł. Damurski, *The Foundations of Multi-channel Neighborhood Governance*, Local and Urban Governance, https://doi.org/10.1007/978-3-031-75124-0_4

Innes and Booher (2010) outline two mental models of public administration: the traditional one, based on a top-down hierarchy of power, and the collaborative one, involving horizontal organization and partnership of participants. The first type of administration narrows the public debate, as public participation is restricted to those who have officially established roles in the process, and its main aim is to raise effectiveness. The second type opens up the process, building and supporting networks of participants. It welcomes all sincere arguments and aims at understanding and agreement (Sager 2009).

Today the traditional model of administration, with its concept of hierarchical bureaucracy and formal decision-making procedures (described as "governing"), is being replaced by the communicative one, focused on inclusion and cooperation, based on horizontal organization (described as "governance") (Innes and Booher 2010; Getimis 2012). When applied to urban governance, the collaborative model involves a broad range of affected groups in a socially oriented and fairness-seeking decision-making process (Albrechts and Denayer 2001). In this approach, development efforts aim to address the interests of large segments of society through consensus-building practices (Sager 2009).

The arguments for collaborative governance may appear attractive to urban policymakers, offering a model that promotes the rules and style that overcome narrow self-interest and incorporate different (also contradictory) views into public deliberation and negotiation. However, those advantages must be confronted with the reality of political power distribution, including democratic accountability within representational politics, transparency in decision-making, and preventing powerful stakeholders from defining the agenda (cf. Scott et al. 2012).

In order to properly evaluate the potential of ICT in neighborhood governance, this visible tension between the traditional and collaborative models needs to be considered taking a wider perspective of public communication and urban discourse. Also, the informal elements of decision-making processes have to be recognized, because not only the official processes contribute to the effectiveness of urban governance.

4.2 Public Communication and Citizen Participation

Public communication in urban governance is a process of exchange of information between all the actors (individuals, groups, institutions, companies) interested in or affected by the development of a particular area. The expected result of communication is a change in the knowledge, opinions, and attitudes of the parties involved (see Pawłowska 2008).

Adopting the principles of public communication to urban policymaking assumes that people seek to reach agreement through mutual understanding, exchange of knowledge, trust, and harmony (Habermas 1999). The traditional instrumental rationality was looking for the best ways to achieve the goals set by the public authorities. The more recent, postmodern trends in public governance aim at achieving communicative rationality, in which the evaluation and selection of goals of local development allows for a nonarbitrary point of view (Belof 2013). The discourse ethics state that communicative process should be open, undistorted, truth-seeking, and empathic: everyone is allowed to express his/her attitudes, desires, and needs but also obliged to reach mutual understanding in dialog (Habermas 1990; Sager 2009).

Over the last five decades, deliberative approaches have moved center stage as a normative model for urban governance. Starting from advocacy planning (Davidoff 1965), followed by the discussion on participatory and collaborative planning (Healey 1997), on empowerment and dissociation (Albrechts and Denayer 2001), and on self-organizing cities (Alfasi and Portugali 2007), and ending up with participatory GIS systems (Sieber 2006) can be summarized with the words of Selle (2005): "planning and development in the city and region is communicating. The vehicle called 'planning' does not move without communication" (Selle 2005, p. 393).

However, the less developed regions still experience particular difficulties in implementing the principles of participatory governance. Political scientists (e.g., Flyvbjerg 1998; Sisk 2001; Juchacz 2006) suggest that the volume of democratic participation is directly connected to the quality of democratic institutions. Representative democracy cannot be perceived as the only way of involving citizens in decision-making processes anymore: a more direct public participation is required. The question arises: how the existing procedures and institutions can be utilized more democratically?

The principles of participatory governance described above have particular implications for the local community level. Neighborhood governance must be transparent and open: the residents have the right not only to know and understand what the authorities are doing but also to actively exercise the power, considering the goals of social cohesion (Somerville et al. 2009).

Therefore, neighborhood governance is a good field to implement novel solutions regarding citizen participation. In particular, the decisions concerning land use can bring visible benefits for democracy—participatory planning ensures that urban structures reflect the needs and preferences of the local community. Among various means of communication, the role of the media may be pointed as an agent supporting the democratization of urban policy.

But communicative, participatory urban governance is not limited to policy issues and the interests of local stakeholders. It is also firmly embedded in urban discourse. An open and transparent discursive environment is a necessary condition for developing democratic policymaking practices. Therefore, in the following section, the concept of urban discourse is presented.

4.3 Embedded in the Urban Discourse

Urban discourse provides a general framework for describing and interpreting urban realities through various forms of public communication and discussion. As an interactive social environment, it helps all participants to understand the contemporary urban ecosystem. However, it also encourages the introduction of new ideas about the future, including the performance of planning tasks and the search for strategic goals. Urban discourse is very flexible and inclusive and can accommodate the perspectives of professionals, stakeholders, and residents.

There are some established urban discourses constructed in the field of urban planning that have a top-down character, such as the creative city, the smart city, the resilient city, the 15-minute city, and others. These discourses exist globally and are not limited to any national context; some of them of course have particular local specificities, but most of them are initiated by some kind of authorities. And there are also bottom-up urban discourses which are shaped by residents. They are underpinned by daily local interactions and sometimes intersect with urban planning discourse but are usually fragmented and elusive, making them difficult to integrate into urban policy processes (Karsten 2009).

Another approach, promoted by Short (2000), distinguishes three fundamental urban discourses: the authoritarian city, the cosmic city, and the collective city. They conceptualize urban social relations and contrast them with a critical analysis of political forces. The authoritarian city is a discourse related to social aggregation: it involves order and discipline, but also resistance, contestation, and anarchy. In this concept, every city materializes the interplay of different powers in the layout of its streets, the location of functions, and the behavior of its citizens. The second—the cosmic city discourse—presents the urban realm as a religious artifact: various beliefs of the past justified the social hierarchy of urban areas, influenced their spatial structures, and led to a specific interplay between the "cosmic" and the "human." Today, due to the progressive secularization of societies, the city is no longer a place of cosmic narrative. And the last concept, collective urban discourse, seems to be the most appropriate here: it presents the city as a site of collective provision and consumption, as well as the practices of civil society. In this approach, the city becomes a shared space, a place of common projects and neighborhood effects (Short 2000).

In this chapter, a simplified definition of an "urban discourse" is adopted. It is understood as a "discussion on urban affairs in speech or writing, producing particular meaning and involving various actors" (Damurski et al. 2023, p. 2). The primary sources of information in this discourse are the media (cf. Talbot 2007), and the quality of communication greatly depends on their ability to create a public forum for exchange of knowledge and opinions.

A relevant example of this relationship is the "local news poverty"—a phenomenon observed in rural areas, suburban municipalities, and smaller towns in Canada. When the local media cannot fulfil the information needs of the communities, local news poverty affects residents with limited access to timely, verified news about local politics, education, health, and other topics (Lindgren et al. 2017). As a result, urban discourse cannot be formed nor developed.

Urban discourse is also an important forum of presenting urban transformation concepts. According to Ossowicz (2019), local media fulfil two specific tasks in shaping urban policy. First, they transfer the information about urban transformation processes to the wider audience; second, they comment on these processes providing mixed expert opinions about the local policy directions. And of course, professional journalism should clearly distinguish those two functions: a supposedly objective presentation of facts builds the core content of the public urban discourse, providing a reliable basis for social interaction; commenting on those facts from many different perspectives (called media framing) means attaching particular weights to given issues, thus influencing people's attitudes. Hence, local media have the power to contribute (directly or indirectly) to the success or failure of housing estates, shopping centers, public spaces, and other types of urban constructions by shaping the stakeholders' views, inspiring them, and persuading to take action (Ossowicz 2019).

Practitioner's Perspective
The impact of the Internet on the quality of information provided by the journalists

The internet has speeded up some processes and thus the reliability dropped. Today I have an impression that unfortunately people read less, and first of all they want to read and listen shortly—an information should last 2–3 minutes no matter if it is to be read or viewed, so this automatically shallows the information.

Anonymous journalists from Poland [interviewees PL06 and PL03 interviewed in 2022; origin: H2020 DEMOTEC research project 2022]

Hence, it can be easily proven that the quality of urban discourse depends on the range and depth of information provided by the media. The accountability of journalists is very high then: democratization of urban communities and increase in civic society values (such as empathy, attentiveness, sensitivity, and tolerance but also accuracy, integrity, and honesty) can be achieved only if people have access to appropriate information and knowledge (Damurski et al. 2023).

In this context, several aspects have to be addressed, including (i) virtualization which has a profound impact on urban discourse as it largely changes the ways people communicate with each other and (2) some informal, bottom-up initiatives which support the deliberative turn in urban policy, becoming a necessary element of building communicative, open urban governance. I will elaborate on each of them below, starting with the latter one.

4.4 The Growing Role of Informality

As it has been already pointed, contemporary urban governance becomes a deliberative system, involving various forms of public communication. Encouraging political activity beyond the state directs a greater emphasis on the role of local communities in policy formulation and implementation (Swyngedouw 2010). However, it also bears particular challenges, one of which is finding a balance of the formal, semiformal, and informal phenomena (cf. Horelli 2013b).

Formality of particular actions is connected with the degree to which a single policy initiative is embedded in the law. This may vary depending on the particular political setting, and a precise definition of what is "formal" and what is "informal" in governance may be difficult. In order to reach at least some general conceptual order, the concept of formality needs to consider the fact that what is seen as formal and informal depends on the criteria adopted in a given location (Altrock 2012). In the hitherto section, I will present the key discussion on this.

First, we must agree that formal processes focus on decisions initiated by public institutions, regulated by the law, with particular people in charge (Damurski 2015). A common understanding of "formal" is then associated with rule-based, structured, predictable, and regular activities undertaken by official public organizations. "Informal" is often thought to be spontaneous and unorganized, but taking a closer look, it is rather simply inspired beyond the institutions and not regulated by the law. Informal activities often generate high levels of public concern (including public protests, marches, media events) as well as lobbying of elected representatives to act in favor of the community's interests (Scott et al. 2012). It is worth noting that acting informally does not necessarily mean a deviation from the formal rules: informality is different from illegality. It is only a lack of official registration that qualifies a particular phenomenon as informal (Altrock 2012).

A radical concept of the "informal city" proposed by Roy (2012) sees urban development as an embodiment of the people's natural entrepreneurial energies or a grassroots uprising against the bureaucracy. It points to the need to rethink both the very notion of the formal/informal divide and the idea of formality as some kind of norm and informality as a deviation. The desire to move beyond such categorizations is expressed in promoting "hybrid arrangements" which mix formal and informal activities, forming a continuum of formality and informality (McFarlane and Waibel 2012).

Taking a more realistic perspective, we need to note that no political system functions on the basis of formal structures and processes alone (Daniels 2004). The progress in social development resulting in more and more complex community composition and performance is usually associated with building advanced systems of institutional settings where activities traditionally considered as informal are incorporated to the formal sphere (Altrock 2012). The importance of informality is constantly growing, offering solutions which are complementary to formal arrangements, especially in contexts where the state is unable or unwilling to perform some of its responsibilities (McFarlane and Waibel 2012).

In some cases, public authorities take a tactical approach in setting the rules. They can, for example, accept the emerging informal initiatives and consequently formalize them as an adequate response to the problems faced by particular communities. Conversely, they sometimes forbid informal actions because of their counter-political effects (cf. McFarlane and Waibel 2012).

The critique of the formal governance arrangements stems from the observation that the citizens' agency is not limited to organized arenas but evolves on a parallel, disorganized discursive level where local politics is made (Hendriks 2006). The transformation of urban areas is not perceived strictly as a result of official planning activities but rather as an outcome of external and internal pressures generated by different stakeholders. "The traditional government still steers through norms, economic incentives, information, policies and programmes, whereas the newer forms of governance rely more on monitoring, deliberation and self-organisation" ((Horelli and Wallin 2013a) p. 15).

As a result, there is an increasing recognition of new policy networks where political power and influence are acquired outside formal public governance (cf. Hillier 2000). The traditional model of policymaking as a top-down activity directed by the government and supplemented by representational democracy is becoming less relevant to the management of public affairs. Informal practices are increasingly mobilized to influence governance outcomes (Scott et al. 2012) and as such are becoming an indispensable element of the decision-making process in cities.

Moreover, it has been proven that municipalities having a wide range of semiformal and informal governance procedures (like masterplans or sectoral strategies) are better prepared for the periods of instability, when the formal decision-making system is challenged or reformed. However, the unconventional procedures require appropriate skills and knowledge, including effective coordination and management, interdisciplinary personnel, ability to cooperate with various stakeholders, dedicated training, evaluation of the results, and constant improvement (Ciesielski 2018). This "informal capacity" of urban municipalities should be constantly expanded in the near future in order to respond to the challenges of the postmodern urban development.

But should we forget about the formal public governance then? The official structures of urban administration still offer a stable basis for planning, and what is needed today is just linking them with local informal networks in order to achieve a successful local governance system (Jarenko 2013). The responsible involvement of professional politicians in partnership with citizens seems to be necessary to address the complexity of problems, even if establishing working, effective relationships between the informality and formality may be a difficult task in the process of sustainable urban development.

To ultimately clarify the terms of "formality" and "informality" in the context of urban governance, a simple three-degree scale can be proposed, considering the stakeholders in charge of a particular activity (initiators) and the relation to the law (whether the activity is obligatory or not) (Table 4.1). This approach may be useful first to classify various policy initiatives and second to describe the potential flows between various categories (notably from informal to semiformal and formal).

Table 4.1 Formal, semiformal, and informal activities in urban governance, depending on the entity responsible and legal nature

Level of formality	Stakeholder in charge	Relation to the law
Formal activity	Public authorities	Required by the law
Semiformal activity	Public authorities	Not required by the law
Informal activity	Other bodies	Not required by the law

Source: based on (Damurski 2015). Origin: NCN research project 2012–2015

Getting back to the virtualization effects on urban governance, we need to note that online communication may generate, enhance, and support any kind of formal, semiformal, and informal activities. On one hand, this process is beneficial to all the involved stakeholders because it offers additional interaction opportunities. On the other hand, it brings new challenges: anonymization, superficial social relationships, automatization, lack of in-person interaction, polarization, and exclusion. In order to provide a broader overview of those processes, the following section will refer to the effects of virtualization on urban discourse.

4.5 Impact of Online Media on the Urban Discourse

After the crosscutting overview of the recent dominant trends in urban governance, it is the time to demonstrate the impact of the ICT on urban discourse and the resulting transition to virtual environment followed by novel online decision-making practices, patterns, and frameworks. The starting point for doing this is a reflection on the overwhelming growth in the number of Internet users.

The ICT Development Index (IDI), a composite measure of progress in information and communications technology, proves that almost all countries have increased their access to the global network (International Telecommunication Union 2014). Over the past 20 years, the online revolution has had an unprecedented impact on societies around the world, with the number of Internet users rising from 400 million in 2000 to 3.9 billion people (51.2% of the world's population) at the end of 2018 (International Telecommunication Union 2018) (the IDI research program has been discontinued since 2018).

Such exponential growth calls for adequate response in public communication practices. "The public sphere – the domain of publishing, the media and social argument – creates virtual public spaces by facilitating communication. Its structures enable people to exchange ideas and discover possibilities without necessarily being present in the same place at the same time" (The Young Foundation 2010, p. 21).

A turning point in the evolution of public communication came with the introduction of interactive online functionalities named "Web 2.0," opening various publishing opportunities to the users ("self-publishing"). It also facilitated making new contacts, keeping up with friends (social media) and looking through the ever-growing online resources with improved search engines (cf. Lourenço and Costa

2010). In terms of urban discourse, the "Web 2.0" brought very important consequences—it widened the range of available information, enhanced the exchange of knowledge and opinions, and eventually made public communication much easier.

The rise of online self-publishing tools has significantly altered the roles of actors contributing to urban discourse. The constant acceleration of information transmission over the last two decades has had a fundamental impact on urban discourse. Digital media thus facilitate interaction between public actors and citizens, allowing for an up-to-date and effective dialog on public issues. A reflection on this phenomenon, presented by the Knight Commission (2009), led to the vision of "informed communities" that maximize the availability of relevant and credible information and support individuals in their active engagement in the public sphere. This approach remains largely valid today, as online technologies continue to reshape the way citizens communicate, socialize, and deliberate with each other and with the public sector.

However, the growing online share of urban discourse, followed by an increasing fragmentation of the local media landscape, also has some negative effects, most notably polarization and misinformation. The phenomenon of "echo chambers" can illustrate this development. It refers to a situation in which users of social media predominantly encounter opinions that match their own and selectively engage with like-minded others. This pattern, enforced by specific algorithms that classify searched content based on past user activity, limits the diversity of information presented to the user. Instead of forming critical and knowledge-based opinions, it leads to online clustering and polarization (Terren and Borge 2021).

The evaluation of the ICT impact on urban discourse from the perspective journalists studied recently in selected European countries (2023) demonstrates that new challenges arise from the digitalization of media, potentially jeopardizing the very nature of urban discourse. Online media are compromising the quality of information, undermining journalists' authority and gradually eroding the essential feature of urban discourse which is enabling effective public discussion on urban affairs (Damurski et al. 2023).

The impact of digital media can be seen in the reduced reliability of information presented by journalists. The accelerated pace of media production, the prioritization of headlines over in-depth content, and the commercial pressure to maximize audience share all lead to a narrow and one-sided view of the world. Finally, the open, deprofessionalized digital media allow politicians to bypass journalists by communicating directly with the public through social media, further polarizing urban discourse.

This predominantly pessimistic picture of the impact of the ICT on urban discourse requires a deep reflection from all the parties involved. The municipal perspective influencing the discourse with particular legal regulations would not be a good solution: any activity of this kind would undermine the freedom of public communication and eventually worsen the current situation. However, a realistic alternative, still embedded in the digital world, emerges from using online tools in citizen participation.

4.6 Urban Information Management and GIS

Information is the key element in any urban planning process (Laurini 2021). It needs to be acquired from a reliable source, properly documented, stored, and analyzed in order to serve as a basis for decision-making process. All of those activities together may be referred to as an information system for urban governance.

For centuries, information systems were based on paper. They had a limited capacity, were available only to some kind of elite, and once created could not be changed. Today the information is live, ever-changing, and (in most cases) available on demand to everyone. Therefore, designing an information system is not just building the database but involves also all organizational aspects surrounding it (Laurini 2021).

The modern geospatial revolution introduces novel methods and technologies for managing data in the process of effective urban governance. For instance, the decision-support systems (DSS) are computer networks that support one or more actors in their decision-making by analyzing issues and proposing solutions for all kinds of real life problems. Their characteristic features have been outlined by Laurini (2021) as follows: different types of management levels; involvement of individuals as well as groups; coverage of all phases of the decision-making process; flexibility and adaptability over time; user-friendliness for all types of users; effectiveness (accuracy, timeliness, quality); ability to analyze, simulate, and model problems; and access to all types of data.

When applied to urban governance, the DSS gains another key feature—it becomes related to space. Geographic information and all software for managing it named GIS (geographic information systems) are an important subset of the urban decision-support system. It includes several groups of functionalities, such as the acquisition and regular updating of data, followed by the provision of related metadata; controlled modeling for projection and forecasting (including what-if simulations); spatial analysis and cartographic presentation; and data management and stewardship (Laurini 2021).

Another aspect of the urban DSS is providing groupware facilities. The group collaboration means people working together on one project (including, e.g., coauthoring of documents). Particular collaborative activities are discussed and coordinated in order to fulfil given tasks. Therefore, a decision-support system for urban planning is not only based on GIS but needs many other information to be considered and integrated.

The growing role of human- and community-centered approaches has led to the transition of geographic systems from tolls for professionals to ones that are accessible to the general public (Aranda et al. 2023). GIS has a great potential to bridge the expert-driven, technical world with bottom-up knowledge from the citizens. Processes starting from volunteered geographic information down to formally approved and officially introduced online interactive participatory tools reflect the need for capturing and using nonexpert spatial information (Brown and Kyttä 2014). As a result, there has been a progress over the past three decades to develop methods for engaging common people to identify the spatial dimensions of urban landscapes, to collect experience-based spatial knowledge from different stakeholders (Kantola et al. 2023).

Schroeder (1996, p. 28) defines public participation geographic information systems (PPGIS) as "a variety of approaches to make GIS and other spatial decision-making tools available and accessible to all those with a stake in official decisions." Today, PPGIS is a standard practice in planning. Public participation in land use policy formulation is often done through online participatory mapping tools (Kantola et al. 2023).

The engagement depends on the practitioners' drive and willingness, but also on the software itself. Therefore, as technology advances, so do PPGIS, becoming more engaging and easier to use (Aranda et al. 2023). However, participatory online systems are only a part of the effective public communication: they cannot be regarded as "the solution" but rather as one of the elements of the holistic process, which should be applied in combination with other modes of the inclusive planning.

4.7 E-participation

Initially, in the early 2000s, the Internet was used only to provide the society with the contact details of particular public administration offices. Later, the idea of e-administration emerged, aiming to make public services available online, which soon became some kind of expected standard in the public authorities' performance (cf. European Commission 2010). This section will frame the concept of e-participation in urban governance which is another step forward toward the digitalization of the decision-making process.

Various interactive technologies shaping contemporary social life are also transforming the practices of activating citizens in urban governance (Brynskov et al. 2014). "The city of the future will be one that grows, evolves and responds according to the needs of its inhabitants. ICTs can open new opportunities for citizens to more actively shape the future of their cities by sparking new forms of civic participation (…) accessing relevant information and enabling a real-time dialogue in which city administrators and citizens can learn from one another" ((UN-HABITAT 2015), p. 6). In line with this vision, the "communicative turn" in urban governance observed in the 1990s and early 2000s has entered a new phase: e-participation (cf. Falleth and Hansen 2011; Scott et al. 2012).

Practitioner's perspective
The growing role of online citizen participation

They [citizens] have a more meaningful role (…) now than they ever had, because they have a platform to call people out, whether it's politicians or journalists, and there are better ways of being represented. There are better ways of counting things, and there are better ways of polling people that are all coming via the Internet and social media.

Anonymous journalist from the UK [interviewee UK10, interviewed in 2022; origin: H2020 DEMOTEC research project 2022]

All of the currently developed online tools in public administration support directly or indirectly participatory decision-making, be it by publishing current information about public services offered through the Internet or building interactive platforms for citizen engagement and discussion. As Höffken and Kloss (2011) note, contemporary citizen participation is inherent in the "Web 2.0" standard as it involves publishing contents produced by the users.

In line with the concept of the formal-informal continuum described earlier in this chapter, e-participation projects can be initiated and implemented by public authorities (to reduce the administrative costs and increase the transparency and efficiency of the decision-making processes) or stimulated by the citizens and local leaders (with the aim of monitoring politicians, promoting civic values, and increasing users' convenience—see (Parliamentary Office of Science and Technology 2009)). In each of those cases, online participation can take different forms, ranging from chats with the local government representatives, through discussion forums, online surveys, and referendums, to the e-voting systems. In this context, it seems obvious that the Internet has a direct impact on the process of public decision-making.

In the beginning of e-participation epoch, some typical patterns of online communication between the local governments and citizens had been studied in selected Polish cities in 2004–2005 (Przybylska 2007). The findings proved that online participation in public decisions was not possible by that time due to the low level of functionality of municipalities' websites. The lack of key information about the decision-making processes, insufficient transparency, and very poor, hardly available communication functions were the main challenges (Przybylska 2007).

Similar research was carried out in 2009 by Sandoval-Almazan and Gil-Garcia (2012) in over 100 Mexican municipalities. Their observations showed that the local government websites offered a good range of information and administrative services. However, the access to the online interaction tools was very limited (usually consisting only of simple online surveys, the results of which were not published), and there were almost no opportunities for active collaboration between the citizens and politicians. In general, the Internet portals of Mexican municipalities remained at a primary level of development: they provided important information, but they did not guarantee a real two-way public communication (Sandoval-Almazan and Gil-Garcia 2012).

Practitioner's perspective
Online participation appreciated by neighborhood planners

All of our staff has welcomed the addition of virtual outreach events because it has brought us much more participation. We, from staff all the way to political leadership, are committed that no matter what the future looks like, the city of Somerville will never give up the ability to have online channels.

Rebecca Lyn Cooper, Senior Planner, Planning & Zoning Division, Mayor's Office of Strategic Planning & Community Development, Somerville [interview conducted via MS Teams 11/19/2021; reviewed and authorized by the interviewee on 11th of December 2021; origin: Fulbright Senior Award scholarship 2021]

In Germany, an interesting study was performed by Emmer (2010), measuring three indicators of citizens' political activity: access to political information (referring to political parties and their programs), interpersonal communication (contacting politicians as well as discussing political issues with others), and participation (petitions, taking part in various forums related to decision-making). The research showed that there is a moderate but stable effect of political mobilization among citizens generated through the online availability of political information. Surprisingly, at that time (2002–2009), the Internet had not replaced other sources of political information or opportunities for participation. It pointed however to the rise of so-called online generation—the new society that will undoubtedly tend to participate more and more in public affairs via the web (Emmer 2010).

According to a later study by Damurski (2012, 2016), the e-participation tools used in urban planning can be evaluated according to a set of four criteria: transparency (accessibility of information on local spatial policy), spatiality (existence of tools which enable mapping various information related to spatial planning), interactivity (possibilities for active citizen involvement via the Internet), and mobility (special functionalities for mobile devices) (Damurski 2012, 2016). This framework adopts the user's perspective, allowing for the comparison of different governance systems, including both formal and informal phenomena.

Box 4.1: E-participation Tool for Official Urban Planning Procedures in Düsseldorf

Interactive websites enable the exchange of information between users and the system administrator. In the case of public administration, "interactivity" shows the level of openness of the authorities to cooperate with the community in the decision-making process. The range of citizen involvement may be stimulated—among other tools—by the municipal websites if they offer ready fill-in forms useful in official participation procedures and if they publish the results of the past public consultations. In 2012, the municipal webpage of the city of Düsseldorf included an active online form to express the users' opinion about the currently published planning documents ("Flächennutzungsplan" and "Bebauungsplan"). This tool gives an opportunity to type the text comment and enables the upload of an external file (e.g., graphics or drawing). The results of the past consultations are also published online.

(continued)

Box 4.1 (continued)

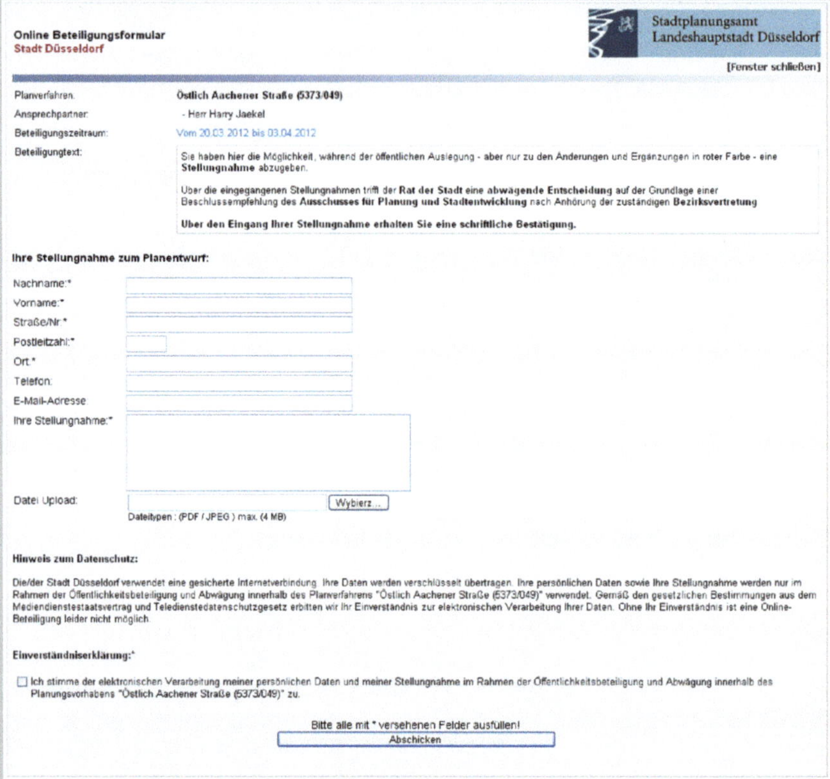

Online Beteiligungsformular
Stadt Düsseldorf

Stadtplanungsamt
Landeshauptstadt Düsseldorf

[Fenster schließen]

Planverfahren: **Östlich Aachener Straße (5373/049)**

Ansprechpartner: - Herr Harry Jaekel

Beteiligungszeitraum: Vom 20.03.2012 bis 03.04.2012

Beteiligungtext:

Sie haben hier die Möglichkeit, während der öffentlichen Auslegung - aber nur zu den Änderungen und Ergänzungen in roter Farbe - eine **Stellungnahme** abzugeben.

Über die eingegangenen Stellungnahmen trifft der **Rat der Stadt** eine **abwägende Entscheidung** auf der Grundlage einer Beschlussempfehlung des **Ausschusses für Planung und Stadtentwicklung** nach Anhörung der zuständigen **Bezirksvertretung**

Über den Eingang Ihrer Stellungnahme erhalten Sie eine schriftliche Bestätigung.

Ihre Stellungnahme zum Planentwurf:

Nachname:*
Vorname:*
Straße/Nr.*
Postleitzahl:*
Ort:*
Telefon:
E-Mail-Adresse:
Ihre Stellungnahme:*

Datei Upload: Wybierz...
Dateitypen : (PDF / JPEG) max. (4 MB)

Hinweis zum Datenschutz:

Die/der Stadt Düsseldorf verwendet eine gesicherte Internetverbindung. Ihre Daten werden verschlüsselt übertragen. Ihre persönlichen Daten sowie Ihre Stellungnahme werden nur im Rahmen der Öffentlichkeitsbeteiligung und Abwägung innerhalb des Planverfahrens "Östlich Aachener Straße (5373/049)" verwendet. Gemäß den gesetzlichen Bestimmungen aus dem Mediendienstestaatsvertrag und Teledienstedatenschutzgesetz erbitten wir Ihr Einverständnis zur elektronischen Verarbeitung Ihrer Daten. Ohne Ihr Einverständnis ist eine Online-Beteiligung leider nicht möglich.

Einverständniserklärung:*

☐ Ich stimme der elektronischen Verarbeitung meiner persönlichen Daten und meiner Stellungnahme im Rahmen der Öffentlichkeitsbeteiligung und Abwägung innerhalb des Planungsvorhabens "Östlich Aachener Straße (5373/049)" zu.

Bitte alle mit * versehenen Felder ausfüllen!

Abschicken

Online participation platform for urban planning in Düsseldorf. (Source: Düsseldorf 2012. Permission for reprint has been implied based the lack of response from the copyright owner. *Origin: NCN research project 2012–2015*)

From a practical point of view, the success of e-participation depends on a number of factors, including the local political culture, the level of social capital, and administrative regulations. Usually e-participation initiatives are perceived as successful when they attract many users, although the importance of certain groups of actors may change in the course of the decision-making process. In order to achieve a higher level of democratization in policymaking, it is necessary to monitor the activity of different users in the subsequent stages of the process and to adopt tailor-made solutions to stimulate the involvement of those who are not interested (cf. Sabo et al. 2011). Another issue is the real impact of e-participation on the decisions taken by the authorities: the question is whether political institutions are able to enter into a dialogical relationship with the public and respect the decisions taken in such dialog (Coleman 2007).

It is symptomatic that prior offline democratic experience is not necessary to participate effectively in online decision-making (cf. De Cindio and Peraboni 2009). However, participation in other online public initiatives is a factor of great importance. And of course, the proper functioning of e-participation tools requires a certain amount of time, as the citizens and administrators have to learn how to use them.

It is impossible to define a universal e-participation instrument that would be suitable for all forms of public consultation. There is a need to develop flexible tools, adaptable to different situations, qualitative rather than quantitative, built in a modular way, allowing various arrangements and functionalities (Lourenço and Costa 2010). The new possibilities offered by the ICT should be successfully incorporated into the practice of public governance, taking into account the changing behaviors, preferences, and communication styles of the subsequent generations of citizens (see De Cindio and Peraboni 2009).

Hence, the key requirements for e-participation can be listed as follows: flexibility, modularity, qualitative character, and strong relation with offline events. These characteristics refer to public decisions in general, but they particularly apply to urban governance. Flexibility seems to be of great importance, regarding the necessity to incorporate semiformal and informal initiatives in urban governance.

The concept of e-participation provides a sound basis for the development of further, more advanced decision-making systems. From a multichannel perspective, e-planning seems to be most relevant, as it enables the collective collaboration of different stakeholders in creating dedicated territorial policies.

4.8 E-planning (Follow-up)

Municipalities increasingly recognize the great potential of the ICT in stimulating urban growth and development and adopt relevant policies to generate the expected benefits for a given city. One of the most influential concepts of contemporary urban governance—the "smart city" (mentioned earlier in this book)—links hard high-tech infrastructures, providing the city's effective daily performance with the availability of high-quality knowledge, public communication, and social capital, perceived as critical factors for a city's competitiveness. It is argued that cities based on the ICT solutions can significantly strengthen their communities and improve the quality of life for all citizens (Leboreiro Amaro 2014).

The strive to become smart may follow multiple directions, depending on the local resources and needs. When adopted to territorial policymaking, it may be labeled e-planning—a term introduced and developed by academics, government agencies, and entrepreneurs since the first decade of the twenty-first century (Klosterman 2012).

Initially the concept was described as a decision-making process supported by ICT systems, aiming to provide a fluent flow of information on the built environment between the users and the planning authorities, with special attention to equal rights of all the actors of the decision-making procedures (cf. Budthimedhee et al. 2002; Szuba 2006). The growing need to develop more specific e-planning intellectual framework was proven by launching the *International Journal of E-Planning Research (IJEPR)* in 2011.

Today's e-planning embraces all the online and offline activities related to planning processes and describes the vital relationships between the ICT and spatial decision-making. It becomes one of the keystones of the current urban governance theory, followed by a growing number of good practices (Damurski 2021). According to some authors, we are witnessing the birth of "digital planners" ("Digitale Urbanisten"—cf. Höffken and Kloss 2011), who not only use the Internet as a useful communication channel in their work but also get actively involved in wider online political debates.

This trend is currently being significantly reinforced by the emergence of artificial intelligence (AI) solutions for urban planning, as an inevitable consequence of technological development which is likely to dominate the next decades of e-planning development. The main purpose of an AI is to translate a human-defined objective into a mathematical one. The outcomes, based on the analysis and abstracting of big data resources can be predictions, recommendations or decisions (Koseki 2022).

As such, AI may play a positive role in improving information acquisition fairness, eliminating spatial and social isolation, as well as reducing information communication barriers. However, it also brings many dilemmas related to the functioning of the democratic decision-making system, enterprises, and the public, urging governments and scholars to respond to the emerging AI challenges (Zhu 2021).

In the urban context, AI systems are essential for the real-time management of the infrastructures and services produced in cities, concerning: traffic management, coordination between health centers, emergency management, mitigating pollution and enhancing predictive maintenance, flexible e-government and e-administration development. The telecommunications network is the backbone of urban infrastructures that supports the correct operation and maintenance of the rest of the urban infrastructures and facilities. In addition, it is responsible for a large part of the social and commercial relations that occur in the city (Ajuriaguerra 2022). Hence, AI affects and changes socioeconomic development patterns and institutional environment that local authorities are working in. It brings new challenges regarding the role of humans in the decision-making process and therefore, in order to avoid long-lasting and widespread failures, it needs to be wisely controlled by human planners. This implies a transformation of urban planning, giving way to a metaverse where the virtual society will complement the physical.

Municipality officials are in a position to shape the enabling environment for the development of AI that in turn supports sustainable and inclusive development. Through the regulatory instruments they can contribute to political accountability referring to the automated urban infrastructural sub-systems, hence balancing the risks and opportunities in developing urban AI (Koseki 2022). A responsible governance and relevant regulatory frameworks will be essential for the success of AI solutions.

Being a part of wider e-administration, supported with AI, e-planning can successfully contribute to territorial policymaking and facilitate active citizen involvement. If a city aims to become "smarter," it should introduce e-planning practices expanding both in-person and in virtual channels of communication (Batty et al. 2012).

The growing popularity of multichannel approaches to territorial policymaking bears particular consequences though. First, personalization of several administrative procedures and adapting them to the needs of individual users is simultaneously accompanied by anonymization of human interaction. Referring to urban services again, we can observe that online channel enables offering dedicated, customized products for particular clients addressing their particular preferences, but at the same time, people using them have no direct face-to-face contact with the retailers which makes the whole process dehumanized (Plesko and Świderski 2015). This observation announces an approaching shift from the massive, uniform thinking about digitalization of urban functions toward an individualized, socially engaged response to specific demands (Liu et al. 2015).

To recapitulate the overview of the e-planning concept presented in this section, I must stress that all new governance paradigms and practices are highly dependent on the availability and usability of ICT infrastructures. If the online connectivity of a given community is low, then the potential of e-planning cannot be fully exploited; similarly, if the planning culture remains authoritarian and formal, then even the best e-planning applications will be useless. This broader sociopolitical and technological context is highly relevant to the way e-planning is adopted and will strongly influence the pace and character of the digital transformation of urban planning (Wallin et al. 2012). All those reflections direct our attention toward wider contexts in which online practices are performed, including the initial idea of digital government and its ancestor—e-governance.

4.9 Digital Government and E-governance

At this point, it becomes clear that the development of ICTs has had a profound impact on the way individuals and organizations think and behave in different social and political settings. Scholars of public policy and administration have paid a lot of attention to the "digital transformation" (Barcevičius et al. 2019), trying to follow the evolution of governance in the last decades. Dawes (2008) presented this brief history as follows: in the early 1990s, despite a significant increase in the availability of computers in the "back offices" of government, most officials communicated by letter, telephone, or simply in-person. Office applications were then slowly introduced, but these were limited to word processing, spreadsheets, and internal e-mail (usually confined to a single building). Communication between different levels of government still relied on the delivery of paper mail (Dawes 2008).

Box 4.2: Incentives to Use the Online Channel

More and more traditional brands develop their online channels and mobile applications. In the process of hybridization, they are trying to convince their clients to start using online ordering with personal pickup. Special promotional materials and advertisements are located all around the stores: on the entrance door, on the shelves, and at the cashier desks. These are followed by appropriate organizational solutions to maximize the "seamless" user experience, including dedicated pickup stands and specialized staff members.

Examples of advertisements encouraging to use online ordering in various shops and restaurants in Boston, MA. (Source: Pictures taken by the author in 2021 and 2022. *Origin: Fulbright Senior Award scholarship 2021*)

Today the global Internet network, laptops, smartphones, and other mobile devices are ubiquitous forms of communication with and within government. Information is often made available for uses beyond the original reason for collecting it. Simultaneously, the boundaries between different institutions and between different levels of government have become permeable due to the ease in transmitting large volumes of information through interconnected, overlapping computer networks, encompassing also nonprofit and private enterprises. Citizens and businesses interact with the government more frequently through e-mail, interactive online forms, and voice systems than in-person or on paper, crossing the boundary between physical and digital realities (Dawes 2008).

Some early debates in this field focused on "digital government" understood as providing the electronic counterpart of traditional administrative operations commissioned by the public sector (Bovaird 2003).

This approach focused on the use of information and technology to support and improve public policies and services, as well as to involve citizens (National Science Foundation 1999). The upgraded definition proposed by the Organisation for Economic Cooperation and Development pointed to the use of the Internet, as a tool to achieve better government (OECD 2003), including improved policy outcomes, higher quality of services, and greater engagement with citizens (OECD 2008). Taken together, these increasingly broad views included both public administration and some elements of the democratic decision-making processes as the main assets of digitalization.

The more recent discussions have introduced the notion of "e-governance" which comprises the use of information and communications technologies not only to support public services delivery but also to develop complex networks of public administration, private sector, civil society, and citizens. As a result, e-governance fosters and strengthens interactive, participatory decision-making processes to solve societal problems collectively (Dawes 2008; Meijer 2015; Gil-Garcia et al. 2018). It can be examined in terms of four interrelated objectives:

- A policy framework where information-related policies play a crucial role in shaping the political agenda
- Enhanced administration performance which embraces customer-oriented delivery of public information
- Constantly monitored, high-quality, and cost-effective government operations including investments in technological infrastructure and introducing organizational innovation
- Citizen engagement which covers a wide spectrum of democratic processes, including accessibility and usability of the ICT in interaction with the authorities, public discourse on political topics, and common agenda-setting

The core arguments in favor of the e-governance lay in the improved efficiency, quality, effectiveness, accountability, and trust either on the demand side or on the supply side (Meijer et al. 2018). Scholars have also adopted a sociotechnical approach to study the digitalization of governance and observed that it goes far beyond its purely technical aspects (Castelnovo and Sorrentino 2018), including also the whole public discourse through which policymakers construct particular social realities (Terlizzi 2021).

Box 4.3: Online Trap in Wałbrzych Participatory Budgeting
In 2014, the Mayor of Wałbrzych City (Poland) introduced participatory budgeting program. Its aim was to allow the citizens to decide how a particular amount of money in municipal budget would be spent. Citizens could propose their own projects which then were put to the public vote for the whole community. In the years 2014–2018, voting took place in both paper and electronic formats. Due to the COVID-19 pandemic restrictions from the 2020 edition onward, participatory budgeting in Wałbrzych has taken place online only. This bears particular issues for seniors and citizens with low technical skills and significantly affects the voting turnout. The municipality is currently adopting solutions supporting the "digitally excluded" groups in participating in the budget, such as involving streetworkers and volunteers equipped with mobile devices enabling online voting.

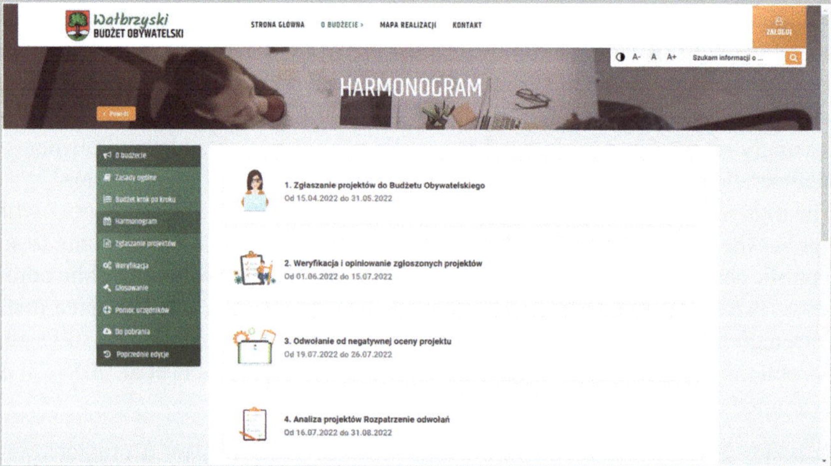

Wałbrzych municipality website dedicated to participatory budgeting. (Source: Wałbrzych 2022. Permission for reprint has been implied based the lack of response from the copyright owner. *Origin: H2020 DEMOTEC research project 2022*)

E-governance issues are more complex and more deeply embedded in social and organizational contexts than the earlier concept of digital government. Online networks enable the exchange of vast amounts of data from a growing number of sources, with multi-scalar impacts on the geographies of governance. Increasing dependence on massive databases has raised security issues, related to the need to protect individuals, organizations, systems, and infrastructure from fraud, errors, hackers, and attacks. The list of other concerns includes public access to data, attracting citizen engagement, as well as information management, use, preservation, sharing, and integration (Dawes 2008).

To recap the discussion presented in this section, it can be noted that e-governance has evolved rapidly from instrumental use of the ICT for standard administrative processes to infusion of technology throughout government offices(Dawes 2008). Municipalities can introduce different e-governance initiatives depending on their current needs and long-term goals; usually they aim at (i) boosting local economic development, (ii) improving the effectiveness of urban service provision, (iii) strengthening social bonds across the local community members, or (iv) promoting the ecological preservation of urban environment (v) collaboration between citizens and public administrations. The multiplicity of e-governance visions shaped within municipal authorities supports the idea that there is no one-size-fits-all approach to digital development (Esposito et al. 2023).

4.10 The Concept of the Local Service Center

A specific tool for multichannel governance has been recently introduced by an interdisciplinary team of researchers directed by Damurski et al. (2020). It is the model of a local service center, which expands the discussion on the hierarchy of urban settlements presented in the previous chapter with novel, digitally wise directions. One of its core assets is incorporating the online channel into neighborhood governance.

The origins of the local service center idea can be located in the ancient Greek polis, where the main city square comprised several buildings of different functions: temples, library, stalls, arsenal, etc. Such a spatial clustering of various activities offered an attractive place for public meetings and for the exchange of goods and thoughts. Later, in the Middle Ages, despite different social, religious, political, and economic contexts, similar functions were performed by relatively small but orderly designed towns with a city square surrounded by buildings offering various facilities (church, market) (Ostrowski 2001). This spatial pattern, continued in the subsequent eras, has crystalized several characteristics of the urban service center: it is a public space offering a specific range of services (retail, administration, finance, religion, culture, entertainment, etc.).

Practitioner's perspective
What could neighborhood planning do to manage both online and offline services?

Zoning is a tool for managing this. In a zoning plan, you can define the types of services that are allowed, and you can also define the way they perform their functions (e.g., the way of shipping products). You can also stimulate the walkability of a neighborhood.

(continued)

Ted Schwartzberg, Neighborhood Planner for Dorchester (north of Park Street), Department of Downtown & Neighborhood Planning, The Boston Planning & Development Agency [interview conducted via Google Meet 10/07/2021; reviewed and authorized by the interviewee on 13th of October 2021; origin: Fulbright Senior Award scholarship 2021]

Our code covers the kinds of services on our streets that don't easily shift to the Internet, so-called experiential services (dentists, doctors, fitness service providers, sandwich shops, small restaurants both for takeout and for dine-in). Through the pandemic, we were able to keep a large percentage of our local businesses because they were all highly desirable services, they're local shops that people need to use (like grocery stores) and that are not easily virtualized, and we were able to move them out to the streets, to public spaces to encourage street life.

Rebecca Lyn Cooper, Senior Planner, Planning & Zoning Division, Mayor's Office of Strategic Planning & Community Development, Somerville [interview conducted via MS Teams 11/19/2021; reviewed and authorized by the interviewee on 11th of December 2021; origin: Fulbright Senior Award scholarship 2021]

Only the industrial era abandoned the idea of the local urban service centers. With its rapid, mass, and initially unplanned urbanization, it disrupted the natural concentration processes and led to the underdevelopment of urban structures. As a result, the role of local center as a structural factor organizing residential zones in urban areas has been lost.

Then, searching for relevant answers to the problems of their epoch, modernists introduced several theoretical concepts aiming to reverse the negative trends in urban development and to restore the multifunctionality of urban settlements. A cornerstone concept at that time was the Perry's Neighborhood Unit (referred already in one of the previous sections of this book). Its goal was to create functional, self-sufficient residential areas with a public square, offices, and a school in the center as well as local stores at the main entrances to the neighborhood (Perry 1929/1998).

The comprehensiveness of the Neighborhood Unit paved the way for later models, one of which was the "modern agora" promoted by Van Klingeren in the 1960s and 1970s. In this approach, the community center was to be a place of social integration, exchange of knowledge, and various citizens. The structure of the agora did not have to be designed for to a specific group of residents—rather it had to be adaptable to the changing needs of its users in various temporal scales (Kowicki 2004).

As it has already been mentioned in this book, in a similar period, starting in 1950s, the American Urban Land Institute developed a typology of shopping centers, based on their size and on the range of services offered. Embedded in the US context, it has precisely distinguished several types of such centers, including the local level where a neighborhood center provides convenience goods (foods, drugs, etc.) and personal services (laundry and dry cleaning, barbering, shoe repairing, etc.) for the daily functioning of the surrounding residential area. It is built around a supermarket, equipped with an on-site parking (Urban Land Institute 2000).

Practitioner's perspective
Supporting the local business with opening online channel

Our economic development colleagues have been very focused on helping the local business owners in getting the resources and expertise needed to get online, to develop a website. Especially during the pandemic, we have done a lot of work with our service providers to set up online ordering and purchasing software.

Rebecca Lyn Cooper, Senior Planner, Planning & Zoning Division, Mayor's Office of Strategic Planning & Community Development, Somerville [interview conducted via MS Teams 11/19/2021; reviewed and authorized by the interviewee on 11th of December 2021; origin: Fulbright Senior Award scholarship 2021]

The local service center is, therefore, an element of the city's spatial structure that can be distinguished from its surroundings due to its specific form and function. Through its human-centered design and spatial arrangement, it supports the daily presence and social interaction of customers. While relatively small in size and focused on satisfying the needs of a housing estate or several face-blocks, it is filled with many key functions, following the natural principle of clustering of human activity. A local service center can be defined as a multifunctional public space providing access to essential daily services, enabling social integration and community-building and promoting local identity (Damurski 2020a, 2020b).

Looking back, the concept arose from the need to create local service centers as an indispensable element of modern urban policy, both at the strategic level (i.e., building a holistic vision of large, multifunctional settlement units) and at the operational level (i.e., implementing specific changes in neighborhoods to improve the quality of life). Five practical guidelines for the creation of local service centers have been recently drawn up, highlighting their main features (Damurski 2020a, 2020b).

The first one points to the need for accessibility and connectivity. Just as the previous examples of local service centers (like the ancient Athenian *agora,* medieval markets, or the Perry's Neighborhood Unit), it must be well connected with other elements of the urban structure. This involves safe walking, easy cycling, and riding other mobile equipment within the local system of public spaces and to other urban locations but also direct access to public transportation network.

The second recommendation is about the local identity. When creating a neighborhood service center, urban policymakers should refer to the history, traditions, and habits of the residents, as well as to the unique values of the place. In other words, supporting locality and place attachment is a core goal, distinguishing a given area from the rest of the city.

The next guideline is connected with the functional and spatial flexibility of the local service center. In order to ensure an optimal number of customers in different parts of the day and in different days of the week, it is necessary to design the neighborhood in such way that it is adaptable to the changing needs and expectations of the residents. This rule involves also the aspects of virtualization, namely, the hybrid services, which use both online and offline channels, are perceived as the best alternative to the strictly physical approach to service provision.

Box 4.4: Multichannel Planning at Somerville Assembly Square

In Sommerville, a city within the Boston, MA, metropolitan area, neighborhood planning took a multichannel approach several years ago. Vibrant public spaces are designed and developed through imposing active land uses at the ground floor, promoting services that can't easily shift to the Internet (dentists, doctors, fitness service providers), supporting local shops that people need to use (sandwich shops, small restaurants both for takeout and for dine-in), and moving them out to the streets to encourage street life. The average for storefronts is 27 feet, which gives spaces mostly suitable for small and local businesses.

Car-free pedestrian area in Assembly Square neighborhood, Somerville, MA. (Source: Picture taken by the author in October 2021. *Origin: Fulbright Senior Award scholarship 2021*)

Guideline number 4 aims to create strong territorial community, in which close personal ties between residents are accompanied with high engagement in decision-making processes. Also here implementing multichannel solutions for social integration and public participation seems to be necessary to secure relatively sustainable neighborhood future.

Eventually the last guideline refers to the wider concept of territorial cohesion, stressing the need to develop local services as a crucial factor in building the overall harmony in human settlements. The well-established hierarchical system of service centers is now being complemented with a functional, horizontal network of interconnected local service centers (Damurski 2020a, 2020b).

4.11 Intermediate Conclusions

The critical review of urban governance paradigms presented at the beginning of this book indicated that new technologies should support planning and management practices. This chapter addressed various aspects of contemporary urban governance in the context of the ever-widening adaptation of ICT solutions in public administration. The main topic in this discussion is the communicative paradigm, pointing to the need for mutual understanding between all the stakeholders in the decision-making process.

All the urban policy activities cofound urban discourse—a public communication environment in which all the local actors exchange information, create meaning, and comment all the urban affairs. One of the main players in the urban discourse is the media. Their role is to convey reliable, in-depth information and enable a debate on urban transformation projects. However, the emergence of online media (including social media) has revolutionized the flow of information in the urban discourse. It is much more challenging to reach and share reliable news now, due to multiplicity of sources of information and the trend of providing short, attractive, and superficial messages.

This affects equally urban development topics, as urban governance is both a subject and an object of public discourse. Despite the apparently easy exchange of information, effective communication becomes more and more difficult and raises particular issues related to digital exclusion of some groups and information overload of the others.

On the other hand, neighborhood governance has all the necessary capacities, competences, and tools at its disposal to build relevant response to online conversion and hybridization of urban services and virtualization of urban life. Its well-established decision-making practices involve a clear territorial reference, economic competences, and effective community engagement using various formal and informal, online and offline channels of communication.

The concept of e-governance presented in this chapter is undoubtedly the most useful one in the search for multichannel urban policy framework. It seems to embrace all the essential aspects of contemporary urban environment: it is embedded in the ICT; it supports participatory decision-making processes and deals with public services. Despite several threats it may bring (like data protection issues, stability of the computer systems, or public access concerns), e-governance is the only reasonable policy concept providing viable solutions to the challenges of virtualization. It will be the core reference for developing the multichannel neighborhood governance at the end of the book.

The examples—both from academic theory and governance practice—presented in this chapter prove that the balance between online and offline environments can be successfully shaped at the neighborhood scale through particular policy solutions:

- Multichannel presence of the neighborhood authorities, effectively communicating with the local community on a daily basis via both social and professional media, including personalized but anonymous messaging with residents
- Online, offline, and hybrid citizen participation options (meetings, debates, workshops, etc.), providing flexibility for formal, semiformal, and informal policy initiatives and inclusiveness for all groups of residents
- Mixed land use in the local service centers (residential buildings, workplaces, public spaces, services and retail)
- Precise regulation of the types of services enabled in the area with special attention to highly desirable, experiential services (healthcare, fitness, barber shops, groceries, bars, cafes, and restaurants) and delivery modes (with preference for in-person pickup)
- Walkable, diverse, and immersive public spaces offering a real alternative to homing and home-centeredness
- Supporting offline service providers in opening their online channels as well as attracting online-native brands to open their offline outlets

All of those ideas and practices will be incorporated in the concept of MCNG presented in the last chapter of this book.

Chapter 5
Conclusions

Abstract This last chapter presents the gaps in the current governance paradigms regarding the relationship between the ICT and urban environment. Several common policy concepts are listed (smart city, urban resilience, neighborhood livability, and others), and the missing components of each of them are pointed out. Then, the research questions asked in the beginning of the book are recalled and answered. The core content of this chapter is the concept of multichannel neighborhood governance, introduced and described as the missing element in the current urban policy puzzle. The last sections reflect on the contribution of the book to the state-of-the-art and propose further research directions.

Keywords Paradigms · Multi-channel approach · Urban governance · Neighborhood governance · Neighborhood services

5.1 Retrospective Overview of the Book

This book aims to capture the neighborhood phenomenon in a possibly comprehensive, holistic way, addressing the recently observed challenges of contemporary urban development. The term "neighborhood" is defined in two ways here: (1) as an administrative unit having particular political power, reflected by dedicated territorial policies, and (2) as a local functional cluster where essential public and private services are provided and where everyday social practices of the local community are performed. Both of those meanings need to be incorporated into the proposed idea of multichannel governance.

In my research, I have tried to present various aspects of the current urban development in the context of the recent rapid growth of the ICT sector. I have focused on the local scale, in line with the notion that neighborhood is the fundamental component of each urban area. Starting with a critical overview of the urban governance paradigms, through a careful analysis of the multifaceted relationships between the Internet and urban functions, followed by a description of selected

Ł. Damurski, *The Foundations of Multi-channel Neighborhood Governance*, Local and Urban Governance, https://doi.org/10.1007/978-3-031-75124-0_5

approaches to local urban spatial policy, I hereby conclude with a concept of multichannel neighborhood governance (MCNG).

The key observation from the studies presented is that the accelerating virtualization of urban life has profound consequences for the spatial, social, and economic structures of cities. At the local level, it is changing the functional profile of neighborhoods, replacing some traditional on-site local services with online ones, reducing (or even eliminating) the need for face-to-face social interaction, rejoining work and home roles, and modifying spatial behaviors of residents. All this calls for a prompt response from the neighborhood governance, which should encompass the aspatial (online) components of the contemporary urban lifestyles.

In this final section of the book, I will recap and synthesize the core findings of the previous chapters. First the gaps in the governance paradigms will be identified and research questions will be addressed, followed by the presentation of the MCNG concept. Eventually a reflection on the contribution of this study and suggestions for further research will be provided.

5.2 Gaps in the Governance Paradigms

From the various urban governance approaches discussed in the previous sections of this book, six seem to be directly relevant for multichannel neighborhood governance: smart city, 15-minute city, urban resilience, neighborhood livability, e-planning, and e-governance. Each of them presents particular values in explaining the online-offline relationships in urban areas and provides some guidelines on how to use ICT in managing urban development. However, none of them addresses explicitly the issue of online conversion of a growing number of urban functions:

1. Smart city focuses on the vital interrelatedness between contemporary urban areas and the IC. However, it provides various concepts, tools, and solutions for the whole city without paying careful attention to the neighborhood level.
2. 15-minute city in its revised version includes also online services and fits very well into the neighborhood scale. It describes how the optimal city should look like and function on daily basis, but does not provide enough measures to achieve this goal.
3. Urban resilience indicates the role of the ICT infrastructures in securing the urban areas and withstanding various disturbances. However, due to its citywide scope, it does not address the particular needs of neighborhood residents in this aspect.
4. Neighborhood livability is an attractive tool for measuring and presenting various aspects of neighborhood performance, including spatial, social, and economic characteristics of particular areas. As such, it gives a useful basis for correcting local policy priorities. Nevertheless, it does not take into account the online layer of a neighborhood which may easily question most of its results.

5. E-planning depicts various ways of managing urban areas by adopting ICT tools to urban governance practices. It provides a comprehensive framework for acquiring, coproducing, processing, publishing, and sharing the data related to public territorial policy. However, it does not address directly the conversion and hybridization of services.
6. E-governance is the most relevant concept for the multichannel approach to urban policy. Its obvious advantages—embeddedness in the ICT, supporting online citizen participation, and facilitating public services delivery—lose their potential at the neighborhood level though. First, e-governance is designed as a decision-making framework for large territories (states, regions, urban areas). It does not take into account the very specific local social, spatial, economic, and political contexts. Second, it focuses on building online solutions as an attractive alternative to traditional governance. It does not seek the vital balance between real and virtual environments, which is necessary to develop holistic and sustainable urban policies. Third, e-governance involves only the public services sector, enhancing its functionalities and increasing its effectiveness. Private sector is not included in its scope, which makes it an incomplete decision-making framework.

The MCNG concept seems to be a missing piece in this policy puzzle. It focuses on both the form and the content of local policy-making and offers solid foundations for the revision of urban governance paradigms.

5.3 Answering the Research Questions and Verifying the Research Thesis

Each study—be it theoretical or empirical—aims to complement the state-of-the-art in a given field with some new knowledge. The questions asked by researchers reflect particular challenges observed in the world which have not been responded before.

In case of this book, the research questions are brought up by the observation of the current urban developments which are opposed to the intellectual patterns followed by urban policymakers. The main queries raised in the introduction were as follows:

- *What are the key challenges presented by the growing presence of ICT in urban environment? How does the virtual reality alter spatial, social, and economic structures of urban areas?*
- *How does urban governance respond to the challenges posed by virtualization? Are the governance paradigms relevant to the current situation at the neighborhood level?*
- *Should online and offline channels be included in neighborhood governance? If yes, what should such multichannel governance involve?*

Responding to all those issues required an in-depth analysis of the current trends in urban development as well as an insight into the urban governance paradigms. This book had an ambition to present each of the two perspectives by reviewing the literature, analyzing selected policy documents and conducting empirical research with policymakers.

As usual, the reality emerged to be much more complex than the theoretical framing. The obtained results provide only partial answers to the research questions. However, some basic conclusions may be drawn regarding each of the issues listed above:

1. ICT brings a profound change in the everyday performance of urban communities and, as a result, impacts the existing spatial structures of urban areas. The main challenges are related to new patterns of socialization (less face-to-face interaction and less social cohesion), work-home relationships (working from home, reduced need for large scale offices), and distribution of goods and services (online ordering and delivery). Many of the traditional urban functions are moving to the Internet. Some will be completely converted and disappear from the urban spaces, while others will be replaced by hybrid solutions that merge online and offline channels. The "customer journey" becomes a seamless experience, where online and offline activities become one continuous process. As a result, the online/offline divide is losing its importance, calling for a holistic, hybrid cognitive framework.
2. Urban governance paradigms consider some aspects of the ICT development and its impact on urban areas. Virtualization is expected to alter the traditional functional structures and policy practices. The common concepts of smart city, 15-minute city, urban resilience, neighborhood livability, or e-planning provide valuable perspectives and solutions to the citywide scale, but they miss the digitalization effects at the neighborhood level.
3. Both online and offline channels are indispensable components of contemporary urban governance. Their role is constantly growing not only in the way the decisions are taken (e.g., online citizen participation), but also in urban development patterns (increasing share of aspatial services in city's functional profile). All those phenomena create the need for a multichannel approach in neighborhood governance, addressing all the current and coming challenges with a comprehensive policy framework.

Those simplistic answers to the research questions need further discussion. In the following sections, I elaborate at length on the arguments and various aspects of the multichannel approach to neighborhood governance. But before doing this, I need also to refer to the core assumption made at the beginning of the book.

The central thesis of this work was that the successful development of urban cores occurring despite the challenges brought by the ICT development cannot be transposed to the neighborhood context and that local, residential urban areas require a careful and dedicated governance approach. Indeed, the advantages of physical proximity which balance the virtualization processes in the citywide scale do not necessarily apply to neighborhoods.

In the citywide scale, various effects of virtualization can be observed. On one hand, the lower costs of communication foster dispersion of those urban functions which can be performed remotely. On the other hand, the need for face-to-face contact supports the centrality advantages. Those trends mutually balance each other, and as a result, core urban areas are unthreatened by online conversion of services.

At the local level, the situation is significantly different. The need for face-to-face contact does not reach particular thresholds to generate enough traffic and thus is not followed by particular economic activity. The scale of the local market is much smaller, and the mix of land uses is much less diverse than in central urban areas, making it difficult to draw the same conclusions as at the citywide scale. Finally, the values attached to space are also different: universality, prestige, and creativity-orientation in central urban nodes versus uniqueness, homeliness, and community building in residential areas. Virtualization supports to some extent each of those values, but is certainly more beneficial to urban cores, contributing to wider polarization effects.

Therefore, the vision of a neighborhood as a place where optimized, diverse, and multichannel small-scale functional structures provide desired living conditions is not a self-fulfilling prophecy. In the times of virtualization, each neighborhood needs a careful, responsive support and guidance from the local authorities: there is a pressing need for conceptual and practical solutions for managing neighborhoods using both online and offline channels.

One more thought to conclude on the main thesis of the book: the two perspectives—citywide and local one—are ultimately interdependent. Even if the core urban areas are doing well in the virtualized world, they have to be backed by residential areas where multichannel lifestyle takes its own forms. If this does not happen, the economic and social polarization within urban areas will grow, annihilating the positive effects in the city centers and downtowns. Only a holistic vision of urban environment, taking advantages of digitalization at all scales, can bring a sustainable, just, and resilient urban futures.

5.4 The Call for a Multichannel Approach in Urban Governance

Problems confronting particular neighborhoods never stand alone but are connected with one another and in most cases also with developments outside the area (Somerville et al. 2009). Considering all the unexpected, tricky changes currently affecting urban areas as a result of virtualization, some may say that only a traditional neighborhood with its social functions, local market, and physical structures could provide appropriate quality of living to local communities. However, I am more than far from defending the status quo of to-date neighborhoods and their governance models. Such defense would make no sense in the world appropriated by metaverse. My aim is to prepare the local governance for changes which are already underway. Planners and policymakers need to internalize several novel facts about the transcending, relational, and hybrid nature of urban neighborhoods in order to find adequate policy solutions.

Traditionally, the spatial and temporal availability of services has been recognized as a key determinant of the quality of life in local communities. Neighborhood livability (meaning an optimized distribution of housing, jobs, and local services that ensures the satisfaction of residents) depends strongly on the accessibility of facilities and on the condition of the local service market (Lovejoy et al. 2010; Mouratidis 2018). Conversely, aspatiality—one of the key characteristics of online services—implies their detachment from localization factors. Service providers may be located anywhere and successfully perform their tasks through the ICT tools. This feature is an antonym of spatiality where geographical location is a binding factor that determines the effectiveness and potential for further progress of traditional urban functions.

The increasing share of aspatial services may change human territorial behaviors and thus reduce the livability of urban areas (cf. Southworth 2016). This is one of the main reasons for developing a multichannel neighborhood governance model.

The demand for a combined, multichannel neighborhood governance has already been discussed in the literature (Damurski et al. 2019). It is based on several observations:

(a) The limited impact of natural concentration mechanisms on the digital world and the resulting virtualization and dispersion of human activities followed by the nonlinearity of agglomeration economies
(b) Conversion of traditional offline services into online ones and the growing role of multichannel (hybrid) providers on the local urban market
(c) The need to redefine the availability, accessibility, and affordability of urban facilities which must now include not only the spatial distribution of services but also their virtual counterparts
(d) The emergence of a relational approach to space, considering territory as a fluid concept, based on its functional profile rather than geographical features; the transient nature of society characterized by high mobility, diversity, and virtualization of interpersonal interactions; and the resulting need to revise the concepts of localism and neighborhood
(e) Ambiguous social effects of conversion and hybridization: online provision of urban facilities focuses on individual needs and reinforces the anonymization of human interaction

In the following section, I will try to satisfy this demand by introducing the concept of MCNG. I will provide a theoretical framework for developing practical solutions for urban governance which consider and tackle the issues listed above.

5.5 The Concept of Multichannel Neighborhood Governance

However trivial this may sound, urban policy is expected to follow the current and anticipate the future trends in social and technological development. This book focuses on the neighborhood as a fundamental component of wider urban structures and attempts to capture the key directions in local communities' evolution associated with the spread of ICT. Its primary aim was to develop a prospective and comprehensive concept of multichannel neighborhood governance as a theoretical framework for managing complex local urban realities in the near future.

The "multichannel" label imposes the requirement of applying online and offline channels to the form and to the content of neighborhood governance. It also enables linking the concept with the ICT sciences which is necessary for finding relevant answers to the emerging virtualization challenges. As Belesky (2020) puts it: "If our disciplines can position themselves as mediators that work between the network layer and the physical layer, we would gain a new way to quantify the performance – and qualify the ambitions – of what we design" (Belesky 2020, p. 102).

Digitalization is a transversal element, cutting through fields, offering attractive governance experience. E-governance comprises the use of information and communications technologies to develop a complex network of various stakeholders and to enhance interactive democratic processes, aiming to solve societal problems collectively. Based on the experience of digital governance practices of the last 20 years and adapting it to the local level, the general definition of multichannel neighborhood governance is as follows:

> ***Multichannel neighborhood governance is a continuous process of decision-making with and for a neighborhood community which includes online and offline channels in both its substance and its procedures.***

In terms of the substance, multichannel governance covers online and offline provision of various urban functions, searching for solutions which promote sustainable, hybrid living environment in the local scale. Neighborhood policy has to acknowledge the multichannel nature of contemporary space, economy, and social relations. Its area of activity includes online and offline constructs: services, distribution of goods, workplaces, human interactions, public spaces, and others.

In its procedures, multichannel neighborhood governance uses online and offline tools to involve all the local stakeholders in the day-to-day management of the neighborhood. It allows bottom-up and top-down initiatives to be combined in a hybrid decision-making process, in which particular solutions are proposed, negotiated, decided, and implemented. It adheres to administrative rules but provides the necessary flexibility in public communication. Stakeholder involvement, citizen participation, and transparency are promoted through the use of both online and offline channels.

In practical terms, MCNG involves various digital decision support systems, which enable online collaboration of different actors, support collecting, and processing of large volumes of data (both spatial and nonspatial, both of human origin and AI-generated) and offer advanced participatory functionalities. Such systems are characterized with high flexibility, user-friendliness, reliability, and transparency. They offer dedicated tools for building decision-making models, attractive interfaces for visualizing geographical information, and simulations of multiple development scenarios. The proposed approach enhances the public management strategies, improves the final quality of local policy decisions but also—in the long run—may substantially contribute to the knowledge society.

Taking a more formal perspective, multichannel neighborhood governance involves both the subjects (people and their powers) and the objects (spatial and functional structures) of the local policy in a hybrid, interactive cooperation framework. It uses all the available means of social communication and welcomes all the local stakeholders to participate in the public decision-making process. It also embraces all the phenomena taking place in the neighborhood, be it material development of urban tissue or virtual development of online functionalities.

Following the discussion on the definition of neighborhood governance presented by Sommerville et al. (2009) and Banfield (2020), MCNG is about the cooperation between different stakeholders, where the shaping and representation of the neighborhood is done jointly by governmental and nongovernmental bodies. While the traditional urban policymaking was conceived in vertical terms, the contemporary neighborhood governance is defined horizontally as a patchwork of spaces over which different civic bodies have responsibility. It emphasizes plurality, bringing together multiple stakeholders with different perspectives to develop neighborhood policies, reduces conflict, and enables responsive reaction to emerging challenges. All this is done simultaneously online and offline, adopting a hybrid view of the concept of neighborhood.

MCNG involves also the organization, supervision, and delivery of goods and services, the maintenance and enforcement of reasonable standards, and provision of acceptable environmental conditions within agreed lines of control and accountability (cf. Power 2004). It cuts across traditional governing patterns and forces local authorities to embrace all the ways in which today's urban citizens fulfil their daily needs, including online and offline channels.

I believe that due to the high pace of the ICT growth and its profound influence on everyday performance of urban areas, policymakers and planners need to take action here and now. The catchy and attractive concept of metaverse will not bring any practical value if it is not followed by particular policy response. As Yoo et al. (2023) put it after (Reynolds 2000), "opportunities for reflection and analysis by academics have been eclipsed by the need for practitioners to take decisions not in regular corporate time but in what has emerged as 'Internet time', rightly discomfited by the rush to action" (Yoo et al. 2023, p. 173).

5.6 Contribution of This Book

Neighborhood governance has a vital role to play in generating economic development and the revival of local communities (ECTP 1998). In order to fulfil this task, its principles have to be regularly revised and updated, adopted to the current civilizational trends. The presented study attempts to critically review the state-of-the-art and propose a novel approach to neighborhood governance, responding to the challenges which have not been addressed by the to-date urban policy.

Conventional urban planning has focused on creating optimal spatial relationships between specific functions located in the urban environment. With the imperative of improving the quality of life of citizens, it has sought to satisfy the preferences of space users, taking into account the anticipated needs of future generations. It has given rise to particular spatial planning rules and practices derived from past development trends and, as a result, has implemented fixed, hierarchical land-use structures that can be observed all over the world (Damurski 2021).

This traditional approach has recently been challenged by the rapid growth of ICT, which has added a dynamic, nonspatial layer to the urban economy. The established patterns and foundations of urban planning have begun to shake, creating the need for a new paradigm. It is no longer sufficient to consider urban development only in geographical terms: it requires including the online dimension in the decision-making process.

In this book, I have tried to find an alternative to conventional planning that would be able to address the virtualization of various urban functions at the neighborhood scale. The review of the current state of knowledge on urban development revealed significant gaps in public governance theory, especially regarding satisfactory solutions for shaping an optimal balance between online and offline environments at the neighborhood level.

The proposed multichannel governance model seems to fill this gap. It embraces all the necessary layers of urban life, and at the same time, it avoids unnecessary interpretative alternatives; it demonstrates a simple and achievable vision of decision-making processes, considering the distribution of powers at the local level; it addresses the current technological development and the resulting urban challenges; it welcomes both formal and informal initiatives in one flexible policy framework.

Box 5.1: Pickup Zones in Poland and in the USA

The growing popularity of online ordering is followed by introduction of relevant spatial solutions. So-called pickup zones become a standard all over the world. Some of them are connected to traditional stores where customers can pick up their shopping personally while others are dedicated to restaurants and cafes where clients wait to be served by the staff. The location rules seem to be quite universal though: close to the store, with easy access to the surrounding streets, appropriate signage, and optimum size proportional to the number of expected clients in the rush hours.

(continued)

Box 5.1 (continued)

Examples of solutions dedicated to fast delivery of food or shopping in Poland and the USA. (Source: Pictures taken by the author in 2021 and 2022. *Origin: Fulbright Senior Award scholarship 2021 and author's own research*)

One of the possible impacts of multichannel neighborhood governance is increasing urban resilience. As it was mentioned earlier in this book, resilience is the ability of an urban system to maintain or rapidly return to desired functions in the face of a disturbance. ICT is a critical component of urban infrastructures supporting the everyday performance of cities and their ability to react to the emerging tensions. If the city in all its spatial scales is equipped with appropriate networks, offering access to necessary urban functions, it is also more resilient.

But resilience is not only about the technical resources and infrastructures that municipalities have at their disposal. It is also about the way they take decisions—what elements they involve and how effective they are in communicating with the stakeholders. In this context, multichannel neighborhood governance has a say. It involves both online and offline channels in decision-making processes and includes both real and virtual layers of urban environment.

Finally, the book also opens new perspectives to urban studies by addressing the issue of aspatiality of online services and its effects to urban governance. Even though it provides only some general recommendations for urban policymakers on how to deal with conversion on the local level, it offers a theoretical basis for further research on the subject.

5.7 Limitations of the Study

There is no doubt that if urban areas are to provide essential services to their citizens and businesses in a continuous, effective way, specific multichannel solutions must be developed and adopted. However, as with any other research, there are a number of caveats to consider.

The first limitation of the presented study (both in its theoretical approach and in empirical evidence) is its geographical scope. The described processes are embedded in the European and US contexts. Hence, they are not representative to the rest of the world. This is one of the reasons why the presented results should be treated with appropriate caution.

The second point is that this book attempts to capture the processes of virtualization that are currently underway. Any study which describes certain phenomena 'in statu nascendi' (in the making) has an unbeatable value: it provides a current picture of the situation in a chosen field of research. Without such attempts, we may miss some momentous shifts in the urban environment and soon find ourselves powerless, adopting policies that are irrelevant to the reality we need to manage. But studying phenomena that are in progress carries also a particular risk—they may turn out to be temporary. It is extremely difficult to assess the reality and propose decision-making frameworks without an extensive, comprehensive, and up-to-date dataset. As a result, the concepts and solutions proposed may soon become outdated which is another reason for an attentive and critical reading, interpretation, and implementation of this work.

Another caveat is connected to the accountability of local authorities. If initiatives such as multichannel neighborhood governance are to work, we need to develop public trust in statutory agencies and faith in the quality of work undertaken by the members of the public (Banfield 2020). Without mutual trust between all the stakeholders, MCNG will become just another inanimate political concept proposed by the academia and never be implemented in practice.

Also the technical issues need to be raised here. Multichannel governance strongly relies on the ICT infrastructures and their functionalities. As one of the activities merging online and offline channels, it is affected by particular threats regarding digital exclusion, security of data, stability of the system, privacy, data stewardship, and costs of maintenance (cf. Dawes 2008). Moreover, it is also threatened by the widespread impact of artificial intelligence, including automation, lack of transparency, fake news distribution or the weakening of professional ethics. Those issues may strongly diminish the applicable potential of the MCNG and should be effectively addressed with further research.

Box 5.2: Mobile Application Supports Displaced Families in Ukraine

The Russian invasion of Ukraine in 2022 caused the fastest forced population movement since World War II. Nearly one-third of the population of Ukraine—more than 14 million people—have fled their homes seeking safe haven inside the country or abroad. Fortunately, a comprehensive government-led system of online services developed before the war provided Ukrainians a lifeline in times of crisis. The application for smartphones launched by the Ministry of Digital Transformation of Ukraine in 2020, known locally as Diia (meaning "action" in English), enables citizens to use digital documents in their mobile devices; it registers and tracks where people are and what situation they are in. It also allows them to directly apply for monthly cash assistance from the state to address the humanitarian needs of their families (Fouani and Brusilovskyy 2022). Until April 2023, it has been downloaded over 10 million times.

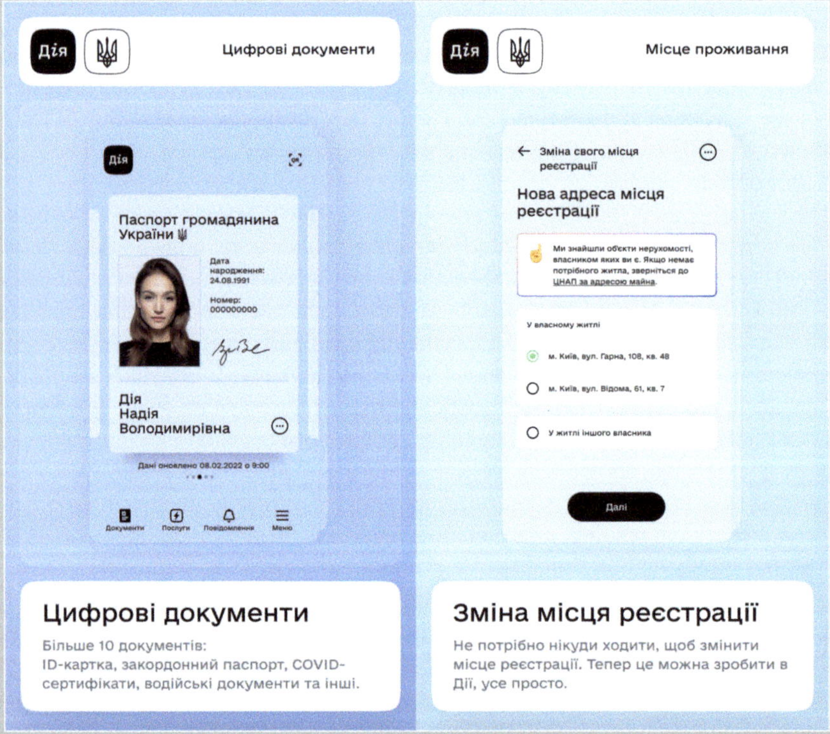

Diia application for smartphones. (Source: Application published by the Ministry of Digital Transformation of Ukraine, available in the Google Play store https://play.google.com/store/search?q=Diia&c=apps&hl=pl&gl=US (accessed 24.04.2023). Permission for reprint has been implied based the lack of response from the copyright owner. Origin: Author's own research)

Lastly, it is necessary to note that good neighborhood governance has always followed the development in human needs and lifestyles, looking far into the future, but never losing the local perspective and the context of political capacities. Its main feature is responsiveness—ability to adapt to the changing human civilization. Without this, multichannel approach will not bring any visible improvement, remaining just an ad hoc reaction to some temporary phenomena. Therefore, the MCNG concept takes an integrative approach, allowing all stakeholders to participate in the decision-making process taking their individual perspectives. This offers a vital mix of viewpoints which is necessary to take optimal, responsive decisions in neighborhoods.

The concept of multichannel neighborhood governance presented in this book paves the way for novel urban policy solutions. But it does not determine their final shape. In order to make it a real-life policy, several steps need to be taken. Operationalization of MCNG will involve promoting it among urban policy practitioners, respecting local political context, matching the existing decision-making practices, and introducing new, multichannel governance procedures. Those tasks will require further preparatory research.

5.8 MCNG Adopted to Services Sector

Much attention has been paid in this book to urban services. They have always been one of the main drivers of urban development, the building blocks of urban spatial structures; they have substantially contributed to the quality of life of individual residents and to the livability of whole communities. At the same time, they are at the forefront of digitalization processes, getting easily converted to online channel and thus supporting the virtualization of the urban environment. I have argued that a novel, multichannel approach is needed to manage this phenomenon. But are omni-channel services manageable at all?

A juxtaposition of the characteristics of single-channel services and multichannel facilities suggests that they are not territorially bound and thus exceed the capacity of conventional urban governance (Damurski 2021). Indeed, it would be difficult to draw a zoning plan for virtual reality where there are neither coordinates nor specified locations. However, as today's economy and society move online, urban spatial policy cannot be left behind. It should follow them by changing its paradigms and methods (Table 5.1).

Thus, omni-channel services are manageable if we look at urban governance beyond its traditional meaning, if we define it as a way of regulating both spatial and nonspatial phenomena, with the aim of improving the quality of life of a given community. The multichannel approach clearly responds to the characteristics of omni-channel services. Consequently, MCNG seems to be a promising policy concept for managing the processes of conversion and hybridization of services—issues that are difficult to grasp but that affect everyday life or urban communities.

Table 5.1 Single-channel versus multichannel approach in urban governance: matching the features of services and urban policy responses

Feature of brick-and-mortar services	**Response of traditional urban governance**
Location that depends on the characteristics of the physical space	Land-use policies that provide attractive locations for businesses
Benefitting from spatial concentration and proximity to customers	Fixed rules and practices aimed at creating a clear hierarchy of service centers
High accessibility requirements	Providing the links between the location of services and transport system
Local customers who usually have a limited influence on the range of the services offered	Lack of capacity to develop supply-demand relationships
Feature of omni-channel services	**Response of multichannel governance**
Relatively low attachment to physical space, resulting in high mobility	Mapping and managing highly dynamic spatial processes, attracting certain types of services that are not easily virtualized
Uneven location patterns and relatively low concentration preferences	Flexible location criteria for services, incentives for online retailers to open their offline outlets
High accessibility requirements for showrooms and collection points	Encouraging multimodal mobility patterns
Different levels of economic activity in online and offline environments	Managing hybrid forms of service provision
Customers from different locations, expecting up-to-date information on products and willing to co-decide on the services offered	Transparent and participatory approach based on highly effective online communication tools, supporting optimal supply-demand relationships

Source: Damurski (2021)

5.9 Further Research Directions

As technology ceases to be seen as something apart from the normal processes of governance and becomes a natural element of a hybrid decision-making framework, it is likely that the division between “online” and “offline” will soon disappear, replaced with more holistic, integrative approaches. The new generation of urban planners and policymakers has just entered the profession—for them, the reality is not divided into the real and the virtual one but forms an unbreakable, hybrid continuum; they use AI everyday for multiple purposes. However, a steady stream of questions regarding the nature and impact of ICTs on public policy remains, driving further research into new directions.

This book presents the rationale and a theoretical basis for developing multichannel approach to local urban governance. The concept of MCNG has been derived from a comprehensive overview of the current civilizational processes and different approaches to local governance. However, this approach needs to be verified in policy practice. Therefore, further research is needed in order to answer the following questions:

- How to operationalize multichannel neighborhood governance to make it implementable in local policy?

- How to overcome the concerns related to artificial intelligence usage, online data protection, privacy, digital exclusion, and system maintenance?
- What local political conditions should be in place to implement the MCNG?
- Which elements of the MCNG concept are universal and which should be adapted to the local context?
- How can the performance of MCNG be evaluated?

Those study directions can be developed in a single, integrated research framework or followed within independent, separate studies. Either way, the core objective is the same: to test the applicable values of multichannel neighborhood governance.

Boston 2021—Wrocław 2024

Bibliography

AARP Public Policy Institute (2021) Livability index. [Online] Available at: https://livabilityindex.aarp.org/. [accessed: 15.06.2021]

Adamski J (2013) Kierunki optymalizacji funkcjonowania usług i infrastruktury kurczących się miast. In: *Zarządzanie rozwojem miast o zmniejszającej się liczbie mieszkańców*. Kancelaria Senatu, Warszawa

Ajuriaguerra M, Abdiu M (2022) Artificial Intelligence in the European Urban Governance. In: J. Ramon Saura, F. Debasa (Eds.) Handbook of Research on Artificial Intelligence in Government Practices and Processes. IGI Global, Hershey, 88–104.

Albrechts L, Denayer W (2001) Communicative planning, emancipatory politics and postmodernism. In: Paddison R (ed) Handbook of urban studies. Sage, London/Thousand Oaks/New Delhi, pp 369–384

Alexander C, Ishikawa S, Silverstein M (1977a) A pattern language. Oxford University Press, New York

Alexander C, Ishikawa S, Silverstein M (1977b) A pattern language: towns, buildings, construction. Oxford University Press, Oxford, UK

Alfasi N, Portugali J (2007) Planning rules for a self-planned city. Planning Theory 6(2):164–182

Allam Z (2020) The first 50 days of covid-19: A detailed chronological timeline and extensive review of literature documenting the pandemic. In: Allam Z (ed) Surveying the Covid-19 pandemic and its implications. Elsevier, Amsterdam, pp 1–7

Allegri A, Ochoa R (2021) Intermittent practices in the contemporary city. The case of Lisbon. In: Advances in human factors in architecture, sustainable urban planning and infrastructure. Springer, Cham, pp 249–256

Allegri A, Ochoa R (2022) Mapping intermittent practices in Lisbon. Human factors in architecture. In: Sustainable Urban Planning and Infrastructure, vol 58, pp 197–204

Altrock U (2012) Conceptualising informality: some thoughts on the way towards generalisation. In: McFarlane C, Waibel M (eds) Urban informalities: reflections on the formal and informal. Ashgate, London

Amazon (2022) Amazon website. [Online] Available at: https://www.amazon.com/gp/help/customer/display.html?ref_=help_search_1-1&nodeId=GQKJHZZQDJBQN2QF&qid=1656943100589&sr=1-1. [accessed: 29.07.2022]

An W, Western B (2019) Social capital in the creation of cultural capital: family structure, neighborhood cohesion, and extracurricular participation. Social Sci Res 81(2019):192–208

Andersen HT, Møller-Jensen L, Engelstoft S (2011) The end of urbanization? Towards a new urban concept or rethinking urbanization. Eur Plan Stud 19(4):595–611

Ł. Damurski, *The Foundations of Multi-channel Neighborhood Governance*, Local and Urban Governance, https://doi.org/10.1007/978-3-031-75124-0

Andrzejewska M et al (2007) O partycypacji społecznej w planowaniu przestrzennym. Zastosowania geowizualizacji w celu wzmocnienia udziału społecznego w planowaniu przestrzennym. [Online] Available at: http://www.gridw.pl/pspe/movies/O%20partycypacji_spolecznej.pdf [accessed: 19.01.2009]
Antunes P, Sapateiro C, Zurita G, Baloian N (2010) Integrating spatial data and decision models in an E-planning tool. Lecture Notes in Computer Science 6257:97–112
Aranda NR, Waegemaeker JD, Weghe NV (2023) The evolution of public participation GIS (PPGIS) barriers in spatial planning practice. Appl Geogr 155(6):102940
Arnstein SR (1969) A ladder of citizen participation. JAIP 35(4):216–224
Arundel R, Ronald R (2017) The role of urban form in sustainability of community: the case of Amsterdam. Environ Planning B Urban Analytics City Sci 44(1):33–53
Bagiński E (1998) Wrocław w opinii swoich mieszkańców. Oficyna Wydawnicza Politechniki Wrocławskiej, Wrocław
Balducci A (2013) Planning the humane smart city. Torino Summer School on Urban studies for a smart city, Torino
Banfield J (2020) Neighbourhood planning. Place, space and politics. Routledge, London/New York
Barcevičius E, Cibaitė G, Codagnone C (2019) Exploring digital government transformation in the EU: analysis of the state of the art and review of literature. EUR 29987 red. Publications Office of the European Union, Luxembourg
Barreda-Ángeles M, Hartmann T (2022) Psychological benefits of using social virtual reality platforms during the covid-19 pandemic: the role of social and spatial presence. Comput Hum Behav 127:107047
Barrera Barrera R, Cepeda Carrión G (2014) Simultaneous measurement of quality in different online services. Serv Ind J 34(2):123–144
Batorski D (2015) Technologie i media w domach i w życiu Polaków. In: Czapiński J, Panek T (eds) Diagnoza Społeczna 2015. Warunki i jakość życia Polaków. Rada Monitoringu Społecznego, Warsaw, pp 373–395
Batty M (2014) City 1.0, City 2.0, City n.0, …, City t. Environ Plann B Plann Des 1(41):1–2
Batty M et al (2012) Smart cities of the future. European Physical J 214:481–518
Belesky P (2020) A perennial practice. Designing between urban landscape and urban network. In: Burry M (ed) Urban futures. Designing the Digitalised City, vol 90, pp 100–107
Bell DR, Gallino S, Moreno A (2014) How to win in an Omnichannel world? MIT SLOAN Manag Rev 5-6(1):44–54
Bell DR, Gallino S, Moreno A (2018) The store is dead – long live the store. Retailing. MIT SLOAN Management Review 2018:1–8
Belof M (2013) Teoria a praktyka planowania regionalnego: doświadczenia pol-skie w planowaniu przestrzennym po 1998 r. Oficyna Wydawnicza Politechniki Wrocławskiej, Wrocław
Bennett L (2017) The many lives of neighbourhood planning in the US: much ado about something? In: Brownill S, Bradley Q (eds) Localism and neighborhood planning. Power to the people? Policy Press, Bristol, Chicago, pp 231–248
Biuro Rozwoju Wrocławia (2019) Wytyczne przestrzenne na rzecz osiedla kompletnego Maślice Małe. Jednostka urbanistyczna E18 w Studium Wrocławia. Wrocław Municipality, Wrocław
Borys T, Rogala P (2008) Jakość życia na poziomie lokalnym—ujęcie wskaźnikowe, Warsaw
Bovaird T (2003) E-government and e-governance: Organisational implications, options and dilemmas. Public Policy and Administration 18(2):37–56
Boysen N, Fedtke S, Schwerdfeger S (2021) Last-mile delivery concepts: a survey from an operational research perspective. OR Spectrum 43:1–58
Bradley Q (2017) Neighbourhoods, communities and the local scale. In: Bradley Q, Brownill S (eds) Localism and neighbourhood planning: power to the people? Policy Press, Bristol, pp 39–55
Brown G, Kyttä M (2014) Key issues and research priorities for public participation GIS (PPGIS): A synthesis based on empirical research. Applied Geography 46:122–136

Brownill S (2017) Neighbourhood planning and the purposes and practices of localism. In: Brownill S, Bradley Q (eds) Localism and neighborhood planning. Power to the people? Policy Press, Bristol, Chicago, pp 19–38
Brownill S, Bradley Q (2017) Reflections on neighbourhood planning: towards a progressive localism. In: Brownill S, Bradley Q (eds) Localism and neighborhood planning. Power to the people? Policy Press, Bristol, Chicago, pp 251–268
Brynskov M, Carvajal Bermúdez JC, Fernández M, Korsgaard H, Mulder I (2014) Urban interaction design: Towards City making. Schloss Neuhausen: UrbanIxD/Booksprints, Amsterdam
Buckner JC (1988) The development of an instrument to measure neighborhood cohesion. Am J Community Psychol 16(6):771–791
Budthimedhee K, Li J, George RV (2002) EPlanning: A snapshot of the literature on using the world wide web in urban planning. J Plann Literature 2002(17)
Byrnjolfsson E, Hu Y, i Raman, M. (2013) Competing in the age of Omnichannel retailing. MIT Sloan Manag Rev 54(4):23–29
Cairncross F (1997) The death of distance: how the communications revolution will change our lives. Harvard Business School Press, Boston
Castells M (1998) The rise of the network society. Wiley-Blackwell, Malden, Oxford
Castells M (1999) The Power of identity. Blackwell Publishing, Oxford
Castelnovo W, Sorrentino M (2018) The digital government imperative: A context-aware perspective. Public Manag Rev 20(5):709–725
Cattell V, Dines N, Gesler W, Curtis S (2008) Mingling, observing, and lingering: everyday public spaces and their implications for Wellbeing and social relations. Health & Place, 14:544–561
Chaskin R (1997) Perspectives on neighborhood and community: A review of the literature. Soc Serv Rev 77(4):521–547
Chaskin RJ (1998) Neighborhood as a unit of planning and action: A heuristic approach. Journal of Planning Literature 13(1):11–30
Checkoway B (1984) Two types of planning in neighborhoods. J Plan Educ Res 3(2):102–109
Chepelianskaia O (2020) Urban resilience: learnings from COVID-19. [Online] Available at: https://www.resilience.org/. [accessed: 12.05.2020]
Christaller W (1933) Die zentralen Orte in Süddeutschland. Gustav Fischer, Jena
Ciesielski M (2018) Urbanistyka poza prawem. Relacje nieformalnych procesów w planowaniu przestrzennym z procedurami skodyfikowanymi. Miasto Pamięć i Przyszłość 3(1):55–63
Clifton J, Díaz-Fuentes D, Fernández-Gutiérrez M (2016) Public infrastructure services in the European union: challenges for territorial cohesion. Reg Stud 50(2):358–373
Coleman S (2007) E-democracy: the history and future of an idea. In: Quah D, Silverstone R, Mansell R, Avgerou C (eds) The Oxford handbook of information and communication technologies. Oxford University Press, Oxford
Colomb C, Santinha G (2014) European Union competition policy and the European territorial cohesion agenda: An impossible reconciliation? State aid rules and public service liberalization through the European spatial planning lens. Eur Plan Stud 22(3):459–480
Colombo M, Mosso C, De Piccoli N (2001) Sense of community and participation in urban contexts. J Community Appl Soc Psychol 11(6):457–464
Cortiñas M, Chocarro R, Elorz M (2019) Omni-channel users and omni-channel customers: a segmentation analysis using distribution services. Spanish J Marketing—ESIC 23(3):415–436
Curwell S et al (2005) Citizens' expectations of information cities: implications for urban planning and design. Build Res Inf 33:55–66
Dąbkowski N, Urbańska W (2012) Budownictwo mieszkaniowe w gminach otaczających wybrane miasta. Kwartalnik Naukowy Problemy Rozwoju Miast II(2012):16–26
Damurski Ł (2012) E-participation in urban planning: online tools for citizen engagement in Poland and in Germany. Int J E-Plann Res 1(3):40–67
Damurski Ł (2015) From formal to semi-formal and informal communication in urban planning. Insights from polish municipalities. Eur Plan Stud 23(8):1568–1587

Damurski Ł (2016) Recent Progress in online communication tools for urban planning: A comparative study of polish and German municipalities. Int J E-Plann Res 5(1):39–54
Damurski Ł (2020a) Five guidelines for local service centres. Prace Komisji Geografii Przemysłu Polskiego Towarzystwa Geograficznego [Studies of the Industrial Geography Commission of the Polish Geographical Society] 34(2):117–128
Damurski Ł (ed) (2020b) Model lokalnego centrum usługowego jako narzędzie kształtowania spójności terytorialnej obszarów miejskich. Oficyna Wydawnicza Politechniki Wrocławskiej, Wrocław
Damurski Ł (2021) How to include Omnichannel services in land-use policy? E-planning holds the key. Int J E-Plann Res 10(3):70–85
Damurski Ł (2022) Neighbourhood cohesion and territorial cohesion: in search for conceptual integrity. GeoJournal 87:4635–4651
Damurski Ł, Ładysz J, Zipser W (2016) Trendy rozwoju usług publicznych i komercyjnych na obszarach podmiejskich na przykładzie powiatu wrocławskiego. Studia Miejskie 22:59–75
Damurski Ł et al (2018) Potential for online conversion of services in urban and suburban neighbourhoods: examples of Wrocław and Siechnice in Poland. In: ATINER's Conference Paper Series, PLA2018–2598, pp 1–12
Damurski Ł, Pluta J, Ładysz J, i Mayer-Wydra, M. (2019a) Online or offline services for urban neighbourhoods? Conceptualisation of research problems. Int J E-Plann Res 8(1):50–67
Damurski Ł, Pluta J, Maier K, Andersen HT (2019b) Stakeholders in the local service Centre: who should be involved in the planning process? Insights from Poland, Czech Republic and Denmark. Bulletin Geog. Socio-economic Series 43:91–106
Damurski Ł, Pluta J, Zipser W (2020) Pedestrian accessibility of services as a measure of territorial cohesion at the neighbourhood level. Bulletin Geo Socio-economic Ser 49(49):31–48
Damurski Ł et al (2023) The impact of digital media on urban discourse: the journalists' perspective. Int J E-Plann Res 12(1):1–20
Daniels PW (2004) Urban challenges: the formal and informal economies in mega-cities. Cities 21(6):501–511
Dao H, Cantoreggi P, i Rousseaux, V. (2017) Operationalizing a contested concept: indicators of territorial cohesion. Eur Plan Stud 25(4):638–660
Davidoff P (1965) Advocacy and pluralism in planning. J Am Inst Plann 31(4):331–338
Dawes SS (2008) The evolution and continuing challenges of e-governance. Public administration review 68(s1):86–S102
De Cindio F, Peraboni C (2009) Fostering e-participation at the urban level: outcomes from a large field experiment "lecture notes in computer science". Electronic Participation 5694:112–124
De Felice F, De Luca C, Di Chiara S, Petrillo A (2023) Physical and digital worlds: implications and opportunities of the metaverse. Procedia Comput Sci 217(2023):1744–1754
De Vreese CH, Boomgaarden H (2006) News, political knowledge and participation: the differential effects of news media exposure on political knowledge and participation. Acta Politica 41(4):317–341
Deakin M, Waer HA (2012) Introduction: from intelligent to smart cities. In: Deakin M, Waer HA (eds) From intelligent to smart cities. Routledge, Oxon, pp 1–7
Denhardt RB, Denhardt JV (2000) The new public service, serving rather than steering. Public Administration Review 60(6):549–559
Dominiak J (2018) Poziom rozwoju nowoczesnych usług w Polsce. Ujęcie regionalne. Prace Komisji Geografii Przemysłu Polskiego Towarzystwa Geograficznego 32(3):204–215
Düsseldorf L (2012) Internet-portal der Landeshauptstadt Düsseldorf. [Online] Available at: http://www.duesseldorf.de/en/index.shtml. [accessed: 29.03.2012]
Dziewoński K (1967) The concept of the urban Economic Base: overlooked aspects. Regional Science Association, Vienna
ECTP (1998) The new charter of Athens 1998. European Council of Town Planners, Athens
ECTP (2003) The new charter of Athens 2003. The European council of town planners' vision for cities in the 21st century. European Council of Town Planners, Lisbon

ECTP (2013) The charter of European planning. The vision for cities and regions – territoires of Europe in the 21st century. European Council of Spatial Planners, Barcelona
Emmer M (2010) Wie hat das Internet die politische Kommunikation der Deutschen verändert? Befunde aus dem DFG-Projekt „Politische Online-Kommunikation" 2002–2009. GESIS-Kolloquium, Mannheim
ESPON (2012) INTERCO. Indicators of territorial cohesion. Scientific platform and tools project 2013/3/2. Final report. Part A. ESPON & University of Geneva, Luxembourg
Esposito G, Terlizzi A, Guarino M, Crutzen N (2023) Interpreting digital governance at the municipal level: evidence from smart city projects in Belgium. Int Rev Adm Sci:1–17
European Commission (2004) A new partnership for cohesion convergence competitiveness cooperation. Third report on economic and social cohesion. European Communities, Brussels
European Commission (2010) Digitizing public Services in Europe: putting ambition into action. 9th benchmark measurement. European Commission, Brussels
European Commission (2011) Cities of tomorrow – challenges, visions, ways forward. Directorate General for Regional Policy, Brussels
European Commission (2013) European innovation partnership on smart cities and communities. In: Strategic implementation plan. European Commission, Brussels
European Union (2011) Territorial agenda of the European Union 2020. Towards an inclusive, smart and sustainable Europe of diverse regions. Gödöllő
Fairgray S (2012) Understanding the geographic relationships between households and retail/services centres across Auckland's urban structure: methodology and regional structure analysis. Auckland Council Technical Report
Falleth EI, Hansen GS (2011) Participation in planning; a study of urban development in Norway. Eur J Spat Develop:1–19
Faludi A (2007) Territorial cohesion and the European model of society. Lincoln Institute of Land Policy, Cambridge
Fischer F (2011) Discursive planning, social justice as discourse. In: Marcuse P et al (eds) Searching for the just city: debates in urban theory and practice. Routledge, London, pp 52–71
Flejterski S, Panasiuk A, Perenc J, Rosa G (2005) Współczesna ekonomika usług. Wydawnictwo Naukowe PWN, Warszawa
Flyvbjerg B (1998) Rationality and power: democracy in practice. The University of Chicago Press, Chicago
Forrest R, Kearns A (2001) Social cohesion, social capital, and the Neighbourhood. Urban Studies 38(12):2125–2143
Fouani M, Brusilovskyy V (2022) How a mobile app to support displaced families in Ukraine is providing a lifeline through the chaos of war. [Online] Available at: https://www.undp.org/ukraine/blog/digital-lifeline-ukrainians-move. [accessed: 24.04.2023]
Francis J, Giles-Corti B, Wood L, Knuiman M (2012) Creating sense of community: the role of public space. Journal of Environmental Psychology 32:401–409
Frysztacki K (1997) Miasta metropolitarne i ich przedmieścia: z problematyki socjologii miasta oraz badań nad rzeczywistością krakowską. Autorów i Wydawców Prac Naukowych Universitas, Kraków
Fukuyama F (2000) Wielki wstrząs: natura ludzka a odbudowa porządku spo-łecznego. Bertelsmann Media, Warsaw
Galster G (2001) On the nature of neighbourhood. Urban Stud 38(12):2111–2124
Galster G et al (2001) Wrestling sprawl to the ground: defining and measuring an elusive concept. Hous Policy Debate 12(4):681–717
Gehl J (2013) Cities for people. Island Press, Washington, DC
Getimis P (2012) Comparing spatial planning systems and planning cultures in Europe. The need for a multi-scalar approach. Plan Pract Res 27(1):25–40
Gil-Garcia JR, Gonzalez-Miranda F (2010) E-government and opportunities for participation: the case of the Mexican state web portals. In: Reddick C (ed) Citizens and e-government: evaluating policy and management. IGI Global, Hershey, pp 56–74

Gil-Garcia JR, Dawes SS, Pardo TA (2018) Digital government and public management research: finding the crossroads. Public Manag Rev 20(5):633–646
Glaeser EL (2010) Agglomeration economics. The University of Chicago Press, Chicago
Goodman A (1977) A comparison of block group and census tract data in a hedonic housing price model. Land Econ 53(4):483–487
Grunig JE, Hunt T (1984) Managing public relations. Harcourt Brace Jovanovich, New York
Gzell S (2013) Praktyka planistyczna a naukowe metody badania miasta. In: Lorens P, Mironowicz I (eds) Wybrane teorie współczesnej urbanistyki. Akapit-DTP, Gdańsk, pp 224–245
Habermas J (1990) Moral consciousness and communicative action. MIT Press, Cambridge
Habermas J (1999) On the pragmatics of communication. Polity Press, Cambridge
Haggett P (1965) Locational analysis in human geography. Butler and Tanner, London
Hajer MA (1993) Discourse coalitions and the institutionalization of practice: the case of acid rain in Britain. In: Fischer F, Forester J (eds) The argumentative turn in policy analysis and planning. Duke University Press, London, pp 43–76
Hajer M, Versteeg W (2005) A decade of discourse analysis of environmental politics: achievements, challenges, perspectives. J Environ Policy Plan 7:175–184
Hall P (2002) Cities of tomorrow. An intellectual history of urban planning and Design in the Twentieth Century, 3rd edn. Malden, Oxford, Victoria
Hallowell I (1955) Culture and experience. University of Pennsylvania Press, Philadelphia
Hartmann T et al (2016) The spatial presence experience scale (SPES). J Media Psychol 28(1):1–15
Harvey D (1989) The condition of Postmodenity. An enquiry into the origins of cultural change. Waley-Blackwell, Cambridge, Oxford
Hastings A (1999) Discourse and urban change: introduction to the special issue. Urban Stud 36:7–12
Häußermann H, Siebel W (1988) Die schrumpfende Stadt und die Stadtsoziologie. In: Soziologische Stadtforschung. Sonderheft
Healey P (1997) Collaborative planning: shaping places in fragmented societies. UBC Press, Vancouver
Helleland B, Ore C-E, Wikstrøm S (2012) Names and identities. Oslo Studies in Language 4(2):117–131
Hendriks C (2006) Integrated deliberation: Reconciling civil society's dual role in deliberative democracy. Political Studies 54(3):486–508
Hillier J (2000) Going round the Back? Complex networks and informal action in local planning processes. Environ Plan A 32:33–54
Höffken S, Kloss S (2011) Digitale Urbanisten oder: Wie das Internet Stadtplanung und urbane Kultur veränder.t [Online] Available at: http://www.vhw.de/fileadmin/user_upload/Forum_Wohneigentum/PDF_Dokumente/2011/FWS_4_2011/FWS_4_11_Hoeffken_Kloss.pdf. [accessed: 26.03.2012]
Hollands RG (2008) Will the real smart city please stand up? Intelligent, progressive or entrepreneurial? City Analysis Urban Trends, Culture, Theory, Policy, Action 12(3):303–320
Horelli L (2013a) Participatory E-planning meets the Glocal. In: Horelli L (ed) New approaches to urban planning. Insights from participatory communities. Aalto University publication series Aalto-ST 10/2013, Unigrafia Oy, Helsinki, pp 131–152
Horelli L (2013b) The contributions of time planning in the Finnish context. In: Horelli L (ed) New approaches to urban planning. Insights from participatory communities. Unigrafia Oy, Helsinki
Horelli L, Wallin S (2010) The future-making assessment approach as a tool for e-planning and community development—the case of ubiquitous Helsinki. In: Silva CN (ed) Handbook of research on e-planning: ICTs for urban development and monitoring. IGI Global, Hershey, pp 58–79
Horelli L, Wallin S (2013a) Starting points. In: Horelli L (ed) New approaches to urban planning. Insights from participatory communities. Aalto University publication series Aalto-ST 10/2013, Unigrafia Oy, Helsinki, pp 11–22

Horelli L, Wallin S (2013b) Towards an architecture of opportunities. In: Horelli L (ed) New approaches to urban planning. Insights from participatory communities. Aalto University publication series Aalto-ST 10/2013, Unigrafia Oy, Helsinki, pp 153–160

Howarth D, Torfing J (2005) Discourse theory in European politics: identity, policy and governance. Palgrave Macmillan, New York

Innes JE, Booher DE (2010) Planning with complexity. An introduction to collaborative rationality for public policy. Routledge, New York

International Telecommunication Union (2014) Measuring the information society report 2014. International Telecommunication Union, Geneva

International Telecommunication Union (2015) ICT Facts & Figures. The world in 2015. International Telecommunication Union, Telecommunication Development Bureau, Geneva

International Telecommunication Union (2018) Measuring the information society report. Volume 1. ITU Publications, Geneva

Jałowiecki B (1989) Rozwój lokalny. Wydział Geografii i Studiów Regionalnych UW, Instytut Gospodarki Przestrzennej, Warsaw

Jałowiecki B, Szczepański MS (2002) Miasto i przestrzeń w perspektywie socjologicznej. Wydawnictwo Naukowe "Scholar", Warszawa

Janc K (2014) Związki pomiędzy światem on-line i off-line na przykładzie widoczności oraz powiązań pomiędzy miastami w cyberprzestrzeni. Studia Regionalne Lokalne 57(3):105–124

Jarenko K (2013) Local co-governance in Herttoniemi: A deliberative system. In: Horelli L (ed) New approaches to urban planning. Insights from participatory communities. Aalto University, Helsinki, pp 45–64

Jones B (1990) Neighborhood planning. A guide for citizens and planners. Planners Press (American Planning Association), Chicago/Washington

Jordan P (2020) Role of place names in relating people and space. In: Handbook of the changing world language map. Springer, Cham

JPI Urban Europe (2015) Transition towards sustainable and Liveable urban futures. The strategic research and innovation agenda of JPI urban Europe. JPI Urban Europe

Juchacz PW (2006) Deliberacja—Demokracja—Partycypacja. Szkice z teorii demokracji ateńskiej i współczesnej. Uniwersytet im. Adama Mickiewicza w Poznaniu, Poznań

Kachniarz M (2012) Efektywność usług publicznych—teoria i praktyka. Wydawnictwo Uniwersytetu Ekonomicznego, Wrocław

Kajdanek K (2012) Suburbanizacja po polsku. Nomos, Kraków

Kantola S, Fagerholm N, Nikula A (2023) Utilization and implementation of PPGIS in land use planning and decision-making from the perspective of organizations. Land Use Policy 127

Kapucu N, Hu Q, Sadiq AA (2023) Building urban infrastructure resilience through network governance. Urban Governanceq

Karsten L (2009) From a top-down to a bottom-up urban discourse: (re) constructing the city in a family-inclusive way. J Hous and the Built Environ 24:317–329

Keller S (1968) The urban neighborhood: A sociological perspective. Random House, New York

Kipta E (2013) Rewitalizacja kurczących się miast, czyli: jak w Polsce realizować hasło don't move, improve. In: Zarządzanie rozwojem miast o zmniejszającej się liczbie mieszkańców (w kontekście perspektywy finansowej 2014–2020). Kancelaria Senatu, Warszawa

Klosterman RE (2012) E-planning: retrospect and Prospect. Int J E-Plann Res 1(1):1–4

Knight Commission on the Information Needs (2009) Informing communities: sustaining democracy in the digital age. The Aspen Institute, Washington, DC

Kooiman J (2005) Governing as governance. Sage, London

Koseki S, Jameson S, Farnadi G, Rolnick D, Régis C, Denis JL (2022) AI and Cities Risks, Applications and Governance. UN Habitat

Kowalewski A et al (2013) Raport o ekonomicznych stratach i społecznych kosztach niekontrolowanej urbanizacji w Polsce, Warszawa

Kowicki M (2004) Współczesna agora. Wybrane problemy kształtowania ośrodków usługowych dla małych społeczności lokalnych. Politechnika Krakowska, Cracow

Kozek W (2011) Gra o jutro usług publicznych w Polsce. Wydawnictwa Uniwersytetu Warszawskiego, Warszawa

Kuhn TS (1970) The structure of scientific revolutions, 2nd edn. University of Chicago Press, Chicago

Kumar V, Anand A, Song, H. (2017) Future of retailer profitability: an organizing framework. Journal of Retailing 93(1):96–119

Kurek S, Wójtowicz M, Gałka J (2013) Przestrzenne zróżnicowanie zmian zaludnienia w Krakowskim Obszarze Metropolitalnym w okresie 1988–2011. Studia Miejskie 12(2013):59–77

Lang T (2010) Urban resilience and new institutional theory – A happy couple for urban and regional studies? In: German Annual of Spatial Research and Policy, pp 15–24

Lang T (2012) How do cities and regions adapt to socio-economic crisis? Towards an institutionalist approach to urban and regional resilience. Raumforschung und Raumordnung 70(4):285–291

Langford M, Higgs G, Jones S (2020) Understanding spatial variations in accessibility to banks using variable floating catchment area techniques. Appl Spatial Anal Policy 3:449–472

Lash S, Urry J (1994) Economies of signs and space. SAGE Publications Ltd, London

Laurini R (2021) Information Systems for Urban Planning: A hypermedia cooperative approach. Taylor&Francis, Boca Raton, https://doi.org/10.1201/9781315274713

Le Corbusier (1973) The Athens charter. With a new foreword by Josep Luis Sert. Grossman Publishers, New York

Leboreiro Amaro A (2014) State of the art of strategic physical planning. In: Arana J, Alonso TF (eds) Strategies for the post-speculative cities. Association of European Schools of Planning, Wrocław

Lee M (1998) Public relations in public administration: a disappearing act in public administration education. Public Relat Rev 24(4):509–520

Lee M (2009) The return of public relations to the public administration curriculum? J Public Aff Educ 15(4):515–533

Leeper TJ, Slothuus R (2020) How the news media persuades: framing effects and beyond. In: Grofman B, Suhay E, Trechsel A (eds) The Oxford handbook of electoral persuasion. Oxford University Press, Oxford, pp 151–168

Lees L (2004) Urban geography: discourse analysis and urban research. Prog Hum Geogr 28:101–107

Leśniak-Moczuk K (2007) Lokalizm w perspektywie globalizacji. [Online] Available at: http://www.univ.rzeszow.pl/ekonomia/zeszyty/Zeszyt8/09_lesniak-moczuk.pdf [accessed: 27.12.2017]

Levin AM, Levin IP, Heath CE (2003) Product category dependent consumer preferences for online and offline shopping features and their influence on multichannel retail alliances. Journal of Electronic Commerce Research 4(3):85–93

Levin AM, Levin IP, Weller JA (2005) A multi-attribute analysis of preferences for online and offline shopping: differences across products, consumers, and shopping stages. J Electron Commer Res 6(4)

Lewicka M (2011) Place attachment: how far have we come in the last 40 years? J Environ Psychol 31(3):207–230

Lewicka M (2012) Psychologia miejsca. Wydawnictwo Naukowe Scholar, Warsaw

Lieber E, Syverson C (2012) Online versus offline competition. In: Peitz M, Waldfogel J (eds) The Oxford handbook of the digital economy. Oxford University Press, New York

Lindgren A, Corbett J, Hodson J (2017) Canada's local news "poverty". Policy options. [online] https://policyoptions.irpp.org/magazines/january-2017/canadas-local-news-poverty/ [accessed 20.03.2025]

Linneman P, Rybczynski W (1999) How to save our shrinking cities. Public Interest 135:30–44

Litman T (2018) Evaluating active transport benefits and costs. Victoria Transport Policy Institute, Victoria

Liu N, Gavino A, Purao S (2015) Extracting citizen values as inputs for designing citizen-responsive urban e-planning services: the VOICE approach and a demonstration in the health-care context. Int J E-Plann Res (IJEPR) 4(2):1–25

Llewelyn-Davies, Banisfter D, Hall P (2004) Transport and City competitiveness: literature review. DfT, London
Location Inc (2021) NeighborhoodScout. [Online] Available at: https://www.neighborhoodscout.com/. [accessed 01.07.2021]
Lorens P (2013) Współczesne trendy zmian w strukturze i funkcjonowaniu miast. In: Lorens P, Mironowicz I (eds) Wybrane teorie współczesnej urbanistyki. Akapit-DTP, Gdańsk
Lourenço RP, Costa JP (2010) E-participation: A discursive approach. Adv Group Decis Negot 1(5):163–184
Lovejoy K, Handy S, Mokhtarian P (2010) Neighborhood satisfaction in suburban versus traditional environments: An evaluation of contributing characteristics in eight California neighborhoods. Landsc Urban Plan 97:37–48
Lusht KM, Farber D (1996) Information technology and urban structure. Real Estate Finance 13:13–20
Lynch PD, Kent RJ, Srinivasan SS (2001) The global internet shopper: evidence from shopping tasks in twelve countries. J Advert Res 41(3):15–23
Maciejuk M (2012) Rozwój budownictwa mieszkaniowego jako przejaw suburbanizacji na Dolnym Śląsku. Studia Miejskie 8:87–94
Madison Reed (2023) Madison Reed website. [Online] Available at: https://www.madison-reed.com/ [accessed: 17.04.2023]
MAM Media (2018) Różnorodność rynkowa w czasach nowego konsumenta. In: Raport z badań ilościowych i jakościowych. MAM Media, Warszawa
Mantey D (2015) Miejsca spotkań w kontekście postmaterialistycznych wartości w suburbiach (przykład gmin Podkowa Leśna i Brwinów). Przestrzeń społeczna 10(2):91–126
Mao L, Nekorchuk D (2013) Measuring spatial accessibility to healthcare for populations with multiple transportation modes. Health and Place 24:115–122
Marcuse P (2015) Depoliticizing urban discourse: how "we" write. Cities 44:152–156
Markowski T (2011) Funkcjonowanie gospodarki przestrzennej—założenia budowy modelu zintegrowanego planowania i zarządzania rozwojem. In: Markowski T, Żuber P (eds) System planowania przestrzennego i jego rola w strategicznym zarządzaniu rozwojem kraju. Komitet Przestrzennego Zagospodarowania Kraju PAN, Warsaw, pp 25–44
Martí P, Serrano-Estrada L, Nolasco-Cirugeda A, López Baeza J (2022) Revisiting the spatial definition of neighborhood boundaries: functional clusters versus administrative neighborhoods. J Urban Technol 29(3):73–94
Martin D (2003) Enacting neighborhood. Urban Geography 24(5):361–385
Martinez-Fernandez C, Audirac I, Fol S, Cunningham-Sabot E (2012) Shrinking cities: urban challenges of globalization. Int J Urban Reg Res 36(2):213–225
McCann P (2003) Geography, trade and growth: problems and possibilities for the New Zealand economy. New Zealand Treasury Working Paper 3(3)
McFarlane C, Waibel M (2012) Introduction: the informal-formal divide in context. In: McFarlane C, Waibel M (eds) Urban informalities: reflections on the formal and informal. Ashgate, London
McKillop T et al (2009) The case for agglomeration economies. Manchester Independent Economic Review, Manchester
Meerow S, Newell JP, Stults M (2016) Defining urban resilience: A review. Landscape and Urban Planning 147:38–49
Meijer A (2015) E-governance innovation: barriers and strategies. Gov Inf Q 32(2):198–206
Meijer A, Bolivar MPR, Gil-Garcia JR (2018) From e-government to digital era governance and beyond: lessons from 15 years of research into information and communications technology in the public sector. J Public Adm Res Theory:1–6
Mills AJ, Plangger K (2015) Social media strategy for online service brands. The Service Industries Journal 35(10):521–536
Ministerial Meeting (2011) Territorial agenda of the European Union 2020. Towards an inclusive, smart and sustainable Europe of diverse regions. Ministerial Meeting of Ministers responsible for Spatial Planning and Territorial Development, Gödöllő

Mironowicz I, Majda T (2017) Manifesty urbanistyczne. IOP Publishing, Towarzystwo Urbanistów Polskich, Warszawa
Mitchell WJ (2000) E-topia: urban life, Jim—but not as we know it. The MIT Press, London–Cambridge
Mitchell WJ (2005) E-topia: information and communication technologies and the transformation of urban life. In: Castells M, Cardoso G (eds) The network society: from knowledge to policy. Johns Hopkins Center for Transatlantic Relations, Washington
Molle W (2007) European cohesion policy. Routledge, London
Moreno A, Barnett J (2021) Glovo: expanding quick commerce. In: Harvard business school, pp 1–30
Moreno C et al (2021) Introducing the "15-Minute City": sustainability, resilience and place identity in future post-pandemic cities. Smart Cities 4:93–111
Mouratidis K (2018) Is compact city livable? The impact of compact versus sprawled neighbourhoods on neighbourhood satisfaction. Urban Studies 55(11):2408–2430
Mullan F, Phillips RL, Kinman EL (2004) Geographic retrofitting: A method of community definition in community-oriented primary care practices. Family Medicine 36(6):440–446
Mumford L (1954) The neighborhood and the neighborhood unit. Town Plan Rev 24(4):256–270
Mumford L (1961) The city in history: its origins, its transformations, and its prospect. Harcourt Brace, New York
Narang S, Reutersward L (2006) Improved governance and sustainable urban development. Strategic planning holds the key. Eur J Spatial Develop:1–11
National Science Foundation (1999) Digital government program announcement NSF. Directorate for Computer and Information Science and Engineering
Neutens T, Schwanen T, Witlox F, De Maeyer P (2010) Equity of urban service delivery: a comparison of different accessibility measures. Environ Plan A 42(7):1613–1635
Nieć M (2010) Komunikowanie społeczne i media. Perspektywa politologiczna. Lex Wolters Kluwer, Warsaw
Nosek Š (2017) Territorial cohesion storylines in 2014–2020 cohesion policy. Eur Plan Stud 25(12):2157–2174
Nowakowski M (1984) Kształtowanie sieci usług. PWN, Warszawa
Nowicki Z, Wojnarowicz P (2015) Efekt ROPO w e-commerce branży RTV/AGD osiąga prawie 80%! In: Kolenda P (ed) Customer journey online. Perspektywy branżowe. Związek Pracodawców Branży Internetowej IAB Polska, Warszawa, pp 36–37
OECD (2003) The hidden threat to E-government: avoiding large government IT failures. Organisation for Economic Co-operation and Development
OECD (2008) The E-government project. Organisation for Economic Cooperation and Development
Ohmer ML et al (2019) Measures for community and neighborhood research. Sage Publications Inc, Los Angeles/London/New Delhi/Singapore/Washington DC/Melbourne
Oleksy T, Wnuk A, Piskorska M (2023) Migration to the metaverse and its predictors: attachment to virtual places and metaverse-related threat. Comput Hum Behav 141:107642
Ossowicz T (2019) Urbanistyka operacyjna. Zarys teorii. Oficyna Wydawnicza Politechniki Wrocławskiej, Wrocław
Ostrowski W (2001) Wprowadzenie do historii budowy miast. Ludzie i środowisko. Oficyna Wydawnicza Politechniki Warszawskiej, Warsaw
Paddison R (2001) Identifying the city: introduction. In: Paddison R (ed) Hadbook of Urban Studies. SAGE Publications, London/Thousand Oaks/New Delhi, pp 11–13
Page BI (1996) Who deliberates?: Mass media in modern democracy. University of Chicago Press, Chicago
Palej A (2005) Poszukiwanie równowagi pomiędzy światem fizycznym a światem wirtualnym we współczesnych miastach. In: Harmensdorfer AD, Cebrat K (eds) Aspects of equilibrium: architecture, urban design, planning at treshold of un decade of education for sustainable development. Oficyna Wydawnicza Politechniki Wrocławskiej, Wrocław, pp 157–163

Palomäki M (1964) The functional centers and areas in south Bothnia. Fennia, Helsinki, p 88
Pandey RU, Garg YK, Bharat A (2013) Understanding qualitative conceptions of livability: An Indian perspective. IJRET: Int J Res Eng Technol 02(12):374–380
Pantano E, Dennis C, De Pietro M (2021) Shopping centers revisited: the interplay between consumers' spontaneous online communications and retail planning. J Retail Consum Serv 61:1–9
Park Y, Rogers GO (2015) Neighborhood planning theory, guidelines, and research: can area, population, and boundary guide conceptual framing? J Plan Lit 1:18–36
Parker G, Salter K, Wargent M (2019) Neighbourhood planning in practice. Lund Humphries, London
Parliamentary Office of Science and Technology (2009) E-democracy. The Parliamentary Office of Science and Technology, London
Pawłowska K (2008) Przeciwdziałanie konfliktom wokół ochrony i kształtowania krajobrazu. Partycypacja społeczna, debata publiczna, negocjacje [Preventing conflicts around conservation and landscape development. Social participation, public debate, negotiation]. Politechnika Krakowska im T. Kościuszki, Kraków
Pendall R, Foster KA, Cowell M (2010) Resilience and regions: building understanding of the metaphor. Cambridge J Reg Econ Soc 3(1):71–84
Perry C (1929/1998) The Neighbourhood unit. Reprinted Routledge/Thoemmes, London
Peterman W (1999) Neighborhood planning and community-based development. The potential and limits of grassroots action. Sage Publications, Thousand Oaks/London/New Delhi
Piercy N, Archer-Brown C (2014) Online service failure and propensity to suspend offline consumption. The Service Industries Journal 34(8):659–676
Plesko R, Świderski S (2015) Żywność. In: Kolenda P (ed) Customer journey online. Perspektywy branżowe. Związek Pracodawców Branży Internetowej IAB Polska, Warszawa, pp 46–48
Polska Izba Handlu (2016) Analiza PIH: Jak podatek od dyskontów i hipermarketów może wzmocnić właścicieli polskich sklepów? Polska Izba Handlu, Warszawa
Pomatto G (2019) Journalists: the role of the media in democratic innovation. In: Elstub S, Escobar O (eds) Handbook of democratic innovation and governance. Edward Elgar, Cheltenham, pp 269–280
Power A (2004) Neighbourhood management and the future of urban areas. Centre for Analysis of Social Exclusion, London School of Economics, London
Przybylska A (2007) Wzory komunikowania w Internecie między samorządem a mieszkańcami. Studia Socjologiczne 3(186):85–114
Reitman AM (2004) What's in a name? The role of naming in neighborhood evolution. Department of Urban Studies and Planning, Massachusetts Institute of Technology, Boston
Reynolds J (2000) eCommerce: a critical review. International Journal of Retail & Distribution Management 28(10):417–444
Richter F (2021) Pandemic accelerates shift to online retail. [Online] Available at: https://www.statista.com/chart/14011/e-commerceshare-of-total-retail-sales/. [accessed: 28.07.2022]
Rohe WM (2009) From local to global. One hundred years of neighborhood planning. Journal of the American Planning Association 75(2):209–230
Roy A (2012) Urban informality. The production of space and practice of planning. In: Weber R, Crane R (eds) Oxford handbook of urban planning. Oxford University Press, New York
Rybczynski W, Linneman P (1999) How to save our shrinking cities. Public Interest 135:30–44
Sabo O, Flak LS, Sein MK (2011) Understanding the dynamics in e-participation initiatives: looking through the genre and stakeholder lenses. Government Information Quarterly 28(2011):416–425
Sack RD (1986) Human territoriality: its theory and history. Cambridge studies in historical geography. CUP Archive, New York
Sager T (2009) Planners' role: torn between dialogical ideals and neo-liberal realities. Eur Plan Stud 17(1)
Sandoval-Almazan R, Gil-Garcia JR (2012) Are government internet portals evolving towards more interaction, participation, and collaboration? Revisiting the rhetoric of e-government among municipalities. Government Information Quarterly 29:72–81

Sanoff H (2000) Community participation methods in design and planning. Wiley, New York/Chichester/Weinheim/Brisbane/Singapore/Toronto

Sassen S (2005) The Global City: introducing a concept. Brown Journal of World Affairs XI(2):27–43

Sassen S (2010) Urban economies and fading distances. In: Buijs S, Tan W, Tunas D (eds) Megacities. Exploring a sustainable future. nai010 Publishers, Rotterdam

Savalia T, Shukla A, Bapi RS (2016) A unified theoretical framework for cognitive sequencing. Front Psychol 7:1821

Scannell L, Gifford R (2010) Defining place attachment: a tripartite organizing framework. Journal of Environmental Psychology 30(1):1–10

Schlappa H, Neill WJV (2013) In: Cities of Tomorrow – Action Today. URBACT II Capitalisation (ed) From crisis to choice: re-imagining the future in shrinking cities. URBACT, Saint-Denis

Schmidt W (2015) Przedmowa. In: Kolenda P (ed) Customer journey online. Perspektywy branżowe. Związek Pracodawców Branży Internetowej IAB Polska, Warszawa, p 1

Schnore LF (1967) Community. In: Smelser NJ (ed) Sociology. Wiley, New York

Schroeder P (1996) Criteria for the design of a GIS2. National Center for Geographic Information and Analysis (NCGIA)

Scott M, Redmond D, i Russell, P. (2012) Active citizenship and local representational politics in twenty-first century Ireland: the role of residents groups within Dublin's planning arena. Eur Plan Stud 20(2):147–170

Selle K (2005) Planen. Steuern. Entwickeln. Über den Beitrag öffentlicher Akteure zur Entwicklung von Stadt und Land. Dortmunder Vertrieb für Bau- und Planungsliteratur, Dortmund

Shankar V, Smith AK, i Rangaswamy, A. (2003) Customer satisfaction and loyalty in online and offline environments. International Journal of Research in Marketing 20(2):153–175

Shearmur R (2010) Scale, distance and embeddedness: knowledge intensive business services location and growth in Canada. In: Doloreux DFM a SR (ed) Knowledge intensive business services: geography and innovation. Ashgate, Farnham Surrey and Burlington Vermont

Shi F et al (2023) A new technology perspective of the Metaverse: its essence, framework and challenges. Digital Communications and Networks

Short JR (2000) Three urban discourses. In: Bridge G, Watson S (eds) A companion to the City. Wiley-Blackwell, New Jersey, pp 18–25

Sieber R (2006) Public participation geographic information systems: A literature review and framework. Ann Assoc Am Geogr 96(3):491–507

Sisk TD (2001) Democracy at the local level. The international IDEA handbook on participation, representation, conflict management, and governance. International IDEA Handbook Series, Halmstad

Skolik S (2015) Zachowania terytorialne w internecie a inflacja przestrzeni społecznej. Przestrzeń społeczna Social space 1:1–16

Smagacz-Poziemska M (2015) Czy miasto jest niepotrzebne? (Nowe) przestrzenie życiowe młodych mieszkańców miasta. Wydawnictwo Naukowe Scholar, Warszawa

Somerville P, Van Beckhoven E, Van Kempen R (2009) The decline and rise of Neighbourhoods: the importance of Neighbourhood governance. European Journal of Housing Policy 9(1):25–44

Southworth M (2016) Learning to make liveable cities. J Urban Des 21(5):570–573

Sowa K (2000) Socjologia. Społeczeństwo. Polityka. Wydawnictwo Wyższej Szkoły Pedagogicznej, Rzeszów

Suttles GD (1972) The social construction of communities. University of Chicago Press, Chicago

Swyngedouw E (2010) City or polis? Profitable politics… or the end of the political. In: Buijs S, Tan W, Tunas D (eds) Megacities. Exploring a sustainable future. 010 Publishers, Rotterdam

Sztompka P (2016) Kapitał społeczny. Teoria przestrzeni międzyludzkiej. Wydawnictwo Znak, Kraków

Szuba B (2006) Wstęp do planowania cyberprzestrzennego. In: Ossowicz T, Zipser T (eds) Urbanistyka w działaniu. Teoria i praktyka. Urbanista, Wrocław

Talbot M (2007) Media discourse: representation and interaction. Edinburgh University Press, Edinburgh

Talen E (2000) The problem with community in planning. Journal of Planning Literature 15(2):171–183

Taylor R, Gottfredson S, Brower S (1984) Neighborhood naming as an index of attachment to place. Population and Environment 7:103–125

Technology (2022) Technology review. [Online] Available at: https://www.technologyreview.com/2018/01/21/241380/amazons-checkoutfree-grocery-store-is-opening-to-the-public/. [accessed: 29.07.2022]

Terlizzi A (2021) The digitalization of the public sector: A systematic literature review. Rivista Italiana di Politiche Pubbliche / Italian Journal of Public Policy 16(1):5–38

Terren L, Borge R (2021) Echo chambers on social media: A systematic review of the literature. Rev Commun Res 9:99–118

The Young Foundation (2010) How can neighbourhoods be understood and defined? The Young Foundation, London/Birmingham

Toffler A (1997) Trzecia fala, PIW, Warsaw

Tuan Y (1977) Space and place: the perspective of experience. University of Minnesota Press, Minneapolis

Tuan YF (1991) Language and the making of place: A narrative-descriptive approach. Annals of the Association of American Geographers 81(4):684–696

Twenge JM (2017) iGen. Why Today's super-connected kids are growing up less rebellious, more tolerant, less happy—and completely unprepared for adulthood. Atria Paperback, New York/London/Toronto/Sydney/New Delhi

UN-HABITAT (2015) The role of ICT in the proposed urban sustainable development goal and the new urban agenda. United Nations Human Settlements Programme, Nairobi

Urban Land Institute (1997a) Dollars & Cents of small town/nonmetropolitan shopping centers: 1997. Urban Land Institute, Washington

Urban Land Institute (1997b) Dollars & Cents of shopping centers in the top 20 metropolitan areas: 1997. Urban Land Institute, Washington

Urban Land Institute (2000) Dollars & Cents of convenience centers: 2000. Urban Land Institute, Washington

Valentini C (2013) Public relations in the public sector. The role of strategic communication in the Italian public administration. I saggi di Sinergie (Papers of Sinergie) 31

Van Neste SL, Sénécal G (2015) Claiming rights to mobility through the right to inhabitance: discursive articulations from civic actors in Montreal. Int J Urban Reg Res 2015:218–233

Vidal-Tomás D (2023) The illusion of the metaverse and meta-economy. Int Rev Financ Anal 86:102560

Wagenaar H (2015) Transforming perspectives: the critical functions of interpretive policy analysis. In: Handbook of critical policy studies. Edward Elgar Publishing, Cheltenham Glos, pp 422–440

Wałbrzych (2022) Wałbrzyski Budżet Obywatelski. [Online] Available at: https://wbo.walbrzych.eu/wszystko-obudzecie/harmonogram,32. [accessed: 28.07.2022]

Wallin S (2013) Urban complexity challenging urban planning. In: Horelli L (ed) New approaches to urban planning. Insights from participatory communities. Aalto University, Helsinki, pp 23–44

Wallin S, Saad-Sulonen J, Amati M, Horelli L (2012) Exploring E-planning practices in different contexts: similarities and differences between Helsinki and Sydney. Int J E-Plann Res 1(3):17–39

Walsh G, Hennig-Thurau T, Sassenberg K, Bornemann D (2010) Does relationship quality matter in e-services? A comparison of online and offline retailing. J Retail Consum Serv 17(2):130–142

Walton J (2000) Urban sociology. In: Quah ASSR (ed) The international handbook of sociology. SAGE Publications, Thousand Oaks

Wang T, Keng-Jung Yeh R, Yen DC, Ari Nugroho C (2016) Electronic and in-person service quality of hybrid services. The Service Industries Journal 36(13–14):638–657

Wdowicka M, Mierzejewska L (2012) Chaos w zagospodarowaniu przestrzennym stref podmiejskich jako efekt braku zintegrowanego systemu planowania (na przykładzie strefy podmiejskiej Poznania). Problemy Rozwoju Miast 1:40–52

Weiss L, Ompad L, Galea S, Vlahov D (2007) Defining neighborhood boundaries for urban Health Research. Am J Prev Med 32(6):154–159

Wellman B, Leighton B (1979) Networks, neighborhoods, and communities: approaches to the study of the community question. Urban Aff Rev 14(3):363–390

Whyte WH (1990) City: rediscovering the center. Doubleday, New York

Wikipedia (2020) Paradigm. [Online] Available at: https://en.wikipedia.org/wiki/Paradigm. [accessed: 22.04.2023]

Williamson W, McFarland P (2012) Investigating the role of electronic planning within planning reform. International Journal of E-Planning Research 1(2):65–78

Wishlade F (2003) Regional state aid and competition policy in the EU. Kluwer Law International, London

Wojnarowska A (2017) Jakość przestrzeni publicznej centrum miasta. Przykład miast średnich regionu łódzkiego. Wydawnictwo Uniwersytetu Łódzkiego, Łódź

Wojtowicz R (1998) Lewis Mumford and American modernism: Eutopian theories for architecture and urban planning. Cambridge University Press, New York

The World Bank Group (2009) Definition of E-Government, World Wide Web. http://www.worldbank.org/egov

Yoo K, Welden R, Hewett K, Haenlein M (2023) The merchants of meta: A research agenda to understand the future of retailing in the metaverse. J Retailng 99(2):173–192

Youssef KWF (2015) Single-access Neighbourhoods and Neighbourhood cohesion. Critical Housing Analysis 2(2):1–10

Zhang X, Mu L (2019) The perceived importance and objective measurement of walkability in the built environment rating. In: Environment and planning B: urban analytics and City science, pp 1–17

Zhang J, Wang T (2023) Urban resilience under the COVID-19 pandemic: a quantitative assessment framework based on system dynamics. Cities 136:104265, 1–20

Zhu WJ (2021) Artificial Intelligence and Urban Governance: Risk Conflict and Strategy Choice. Open Journal of Social Sciences, 9, 250–261. https://doi.org/10.4236/jss.2021.94019

Zhu X et al (2014) A retrospective study on changes in residents' physical activities, social interactions, and neighborhood cohesion after moving to a walkable community. Preventive Medicine 69:93–S97

Zipf KG (1949) Human behavior and the principle of least effort. Addison-Wesley Press, Cambridge

Zipser T (1983) Zasady planowania przestrzennego. Oficyna Wydawnicza Politechniki Wrocławskiej, Wrocław

Index

Ł. Damurski, *The Foundations of Multi-channel Neighborhood Governance*, Local and Urban Governance, https://doi.org/10.1007/978-3-031-75124-0

MIX
Papier aus verantwortungsvollen Quellen
Paper from responsible sources
FSC® C105338

If you have any concerns about our products,
you can contact us on
ProductSafety@springernature.com

In case Publisher is established outside the EU,
the EU authorized representative is:
Springer Nature Customer Service Center GmbH
Europaplatz 3, 69115 Heidelberg, Germany

Printed by Libri Plureos GmbH
in Hamburg, Germany